40 FUNDAMENTALS

of

English Riding

40 FUNDAMENTALS

of

English Riding

Essential Lessons in Riding Right

Hollie H. McNeil

Storey Publishing

*The mission of Storey Publishing is to serve our customers by
publishing practical information that encourages
personal independence in harmony with the environment.*

Edited by Lisa H. Hiley and Deborah Burns
Art direction by Cynthia McFarland and Alethea Morrison
Book design by Alethea Morrison and Cynthia McFarland
Text production by Liseann Karandisecky and Cynthia McFarland

Cover photography by © Carien Schippers
Interior photography by © Carien Schippers, except for © Arnd
    Bronkhorst/www.arnd.nl: v, xii, 48, 81, 116, 150; © Shawn Hamilton/
    CLIX Photography: 30, 160; © Sara Lieser/*The Chronicle of the Horse*: 19;
    © Hollie H. McNeil: 169 middle; © Mike McNeil 171
Illustrations by Joanna Rissanen

Indexed by Susan Olason, Indexes & Knowledge Maps

Storey Publishing
210 MASS MoCA Way
North Adams, MA 01247
www.storey.com

Printed in China by Toppan Leefung Printing Ltd.
10 9 8 7 6 5 4 3 2 1

LIBRARY OF CONGRESS CATALOGING-IN-PUBLICATION DATA

McNeil, Hollie H.
  40 fundamentals of English riding / by Hollie H. McNeil.
      p. cm.
  Includes bibliographical references and index.
  ISBN 978-1-60342-789-0 (hardcover with jacket and dvd : alk. paper)
  1. Horsemanship. I. Title.
SF309.M474 2011
798.2—dc22
                                    2010047764

DEDICATION

This book is dedicated to all of the students who have come to me over the years.

It has been my pleasure and honor to help guide you on your journey. Through your open-mindedness and willingness to listen, watch, and learn, I have been granted the opportunity to share a similar journey: to become a better instructor. A mentor of mine, equestrian education pioneer Jill Hassler Scoop, once remarked that she felt like she should have been the one paying her students because of all that they had taught her. I share her sentiments. From every student I have learned something and from every lesson I understand more.

ACKNOWLEDGMENTS

No book is ever written by a single person, and no author ever works truly alone. This is certainly the truth behind a complex project like this.

My thanks and appreciation start with the folks who kept the farm running on a daily basis. Business and farm duties never skipped a beat as photo and video days gobbled up my time and attention and the writing kept me secluded behind closed doors. Andrea Beukema has been especially amazing. She was able to wear her own hat and many of mine.

My thanks also go to my students and friends who gladly joined in, either to loan their horses or to participate in the photo and video days. Many of them are on the Cast of Characters page at the end of the book.

Thanks also to the wonderful professionals at Storey Publishing who saw the vision for this book-and-DVD combination and jumped enthusiastically on board; especially to Deb Burns and Lisa Hiley, who navigated through this very complex project with grace and calm.

Last and most of all, I must acknowledge my patient, supportive, and understanding family. My children, Alex, Sean, and Max, stood behind me every step of the way, whether helping on shoot days or just being patient with me as I wrote. Whatever was needed, large and small, they have been there to provide.

My mother and father deserve my deep thanks too, for helping me chase my riding dreams from the beginning. They're still doing that for me today.

Without a doubt, my husband, Michael, deserves the greatest acknowledgment there is. He is the visionary, the energy, the mechanic, the tactician, and the unrelenting perfectionist who makes everything, and I do mean everything, possible.

— CONTENTS —

FOREWORD

THE BEST RIDING INSTRUCTORS cannot give all their knowledge during riding lessons. Riding lessons are practice sessions where the rider is guided through a systematic process of learning to use his body correctly, to interact with the horse in a sensible and understandable manner, and to develop the feel for the correct movement of the horse in whatever job he is given to do. It is up to the rider to find additional education through watching, listening, and reading. There are many wonderful books available to the rider, from the "old masters" to more modern tomes, and there are DVDs available on various subjects.

Hollie McNeil has created a unique system. Her combined book and DVD make watching, listening, picture viewing, and reading possible all with one program. She has written an excellent book packed with useful photographs and diagrams. Each section of the book corresponds to a section on the DVD. Whether you read the text first, looking at the pictures, and then go to the corresponding area on the video, or watch the exercises on the video before turning to the book, you will find an all-inclusive learning experience. The video is a marvel, with action shown from angles not usually seen and in slow motion as well as at normal speed. The student will want to go through the video of each movement many times to catch everything.

Hollie has done a masterful job explaining the basics and the details in a way that will help any rider. This is a systematic program that gives a real chance for success with most horses. Her 40 Fundamentals are from the classical system of riding but are presented in a clear, understandable way. She also brings in new scientific studies and explains how these studies relate to modern equitation or how we might think differently than the old masters.

This book will be a valuable review for instructors who want to add to their tools for teaching. Teachers will find new ways of expressing themselves and perhaps useful exercises for their lessons. The video gives newer instructors a wonderful opportunity to develop their eye, with the slow-motion and freeze-frame sequences being especially valuable. *40 Fundamentals of English Riding* is a valuable leap forward in helping riders to be the best that they can be.

— Lendon Gray
Bedford, New York

PREFACE

EVERY AUTHOR HAS TO HAVE A MOTIVATION for sitting down to write something. My motivation for *40 Fundamentals of English Riding* was born years ago, when I began to teach riding. Over the years, I have met hundreds upon hundreds of new students. Some come to me without any preconceived notions of what correct riding really is. Others come after taking lessons elsewhere and trying this and that. Still others come after a long hiatus, often after having kids and a career.

What I frequently hear from students who have ridden elsewhere is a sense of suspicion about what I am saying and why. I hear things like "that's not the way I learned," or "my old instructor always had me do it this way." Often, the debate is not over some esoteric aspect of riding theory, but fundamentals like how to hold the reins, how to sit the trot, or the leg aids for canter.

Practically speaking, there is plenty of wiggle room for personal preferences, and those are easy to teach around. Accepting a slightly different approach is simply a matter of broadening your view on any given subject.

The real challenge for me was being put on the defensive and made to feel like I was in a courtroom, trying to win a case. I needed evidence, case citations, and precedent for my assertions.

As a voracious reader of all things equine, I used authorities from around the world to back up my teaching: "According to Conrad Schumacher," or "Kyra Kirklund says," and "Sally Swift teaches that," and so on. (I have often joked that I don't have an original thought in my brain; everything I know is just regurgitated information from others.)

In defense of my personal riding and training theory, I found I became a much stronger teacher because of my need to have evidence and authority to back up my statements.

WE NEED A SOLID UNDERSTANDING OF THE FUNDAMENTALS. As an instructor, I've witnessed a huge void in the understanding of foundation level riding. The need for riders to understand the basics of riding, in an orderly approach, is the genesis for *40 Fundamentals of English Riding*.

I'm not the only one who feels this way. When someone asked Anne Gribbons (United States Equestrian Federation [USEF] dressage technical advisor; Fédération Equestre Internationale [FEI] 'O' Judge; and Pan Am silver medalist) what she would do to improve dressage in the United States, her answer was, "Basics, basics, basics! Riders need to learn the system, learn to pay their dues on the lunge line, to not let overwhelming ambition cloud their judgment."

Highly regarded German trainer Christoph Hess agrees: "Every rider, even Grand Prix, needs to focus on the basics."

The underlying reason for this predicament is that the United States does not have a system in place to educate riders and there are no national standards. Pretty much anyone who wants to teach riding or train horses in this country can do it, regardless of their education, qualifications, or credentials. When a new student walks into a barn for lessons, who knows what information (or misinformation) she will receive?

The lack of standards and consistency can be blamed partly on the size and diversity of the United States. We have millions of riders and millions of horses. And all these horse enthusiasts are not all playing the same game with the same rules.

Even within a single discipline, riders disagree on the basics. The geographically small and densely populated countries of Europe can monitor their horse industries more closely and therefore can guide the education of those participating.

THE FUNDAMENTALS APPLY TO EVERY RIDING DISCIPLINE. People often call our barn for lessons and want to know what kind of barn we are. They ask if we go to hunter/jumper shows; do we have eventers; are we strictly a dressage barn?

This sense of division between the English disciplines — all of which have the same fundamentals — contributes to the lack of cohesive information and organization.

When I was looking into certification programs, I realized that within the United States, I could be certified as a dressage instructor through the United States Dressage Federation (USDF), as a combined training instructor through the United States Eventing Association (USEA), and as a hunter/jumper trainer with the United States Hunter Jumper Association (USHJA). Are these really such different ways of riding and training that I would need a certification for each?

Instead, I went to Germany to earn a single certification license that recognizes my ability in the areas of dressage, jumping, and cross-country, in addition to lungeing, stable management, and horse health. Through this certification I hold an international instructor's passport that is recognized by the Fédération Equestre Internationale (FEI).

I firmly believe that riding should be based on the same principles whether you are heading for the jumper ring, the dressage ring, or the cross-country course. Classical riding and training is the basis for *all* English equestrian sports.

Dressage might sound fancy, but it's simply a French word for the training of the horse. With that understood, it's easy to see that dressage is for *every* horse. The goal of dressage, and therefore of all types of riding, is to make the horse more "rideable."

THE FUNDAMENTALS ARE GOOD FOR EVERY HORSE

Not every horse is headed for the big leagues. I've heard people smirk when looking at a horse that's not a warmblood, saying, "That's not a dressage horse." Perhaps the horse will never earn honors in competitive dressage, but without a doubt, dressage is good for *that* horse.

One little mare who came to my barn had been ridden Western and exclusively on the trail all her life. You would assume she'd be calm and easy to deal with. On the contrary, she wouldn't stand to be mounted, shifted into race mode once you were on, and danced and pranced for the entire ride.

Over the course of several years of dressage riding, however, she has learned to stand patiently while her rider mounts and adjusts stirrups and girth, and then warms up in a relaxed, swinging frame. In a schooling show at training level she even earned a score of more than 70 percent from a pretty tough judge.

Another example of why dressage is good for every horse is the sheer physical therapy that comes with working a horse properly. We had a 15-year-old horse at our barn who was riddled with osteochondrosis (OCD – a breakdown of cartilage that can lead to joint erosion). Her owner brought her in just out of curiosity to see what she might be able to do. There was no doubt this horse had a little "hitch in her giddy-up," but we started working on the basics.

Over time, the mare became more balanced, and more supple and useful with her body. She eventually showed at recognized dressage shows to First Level and earned quite respectable scores. Did dressage give this horse a new lease on life? Absolutely.

RIDING IS AN ART AND A SCIENCE. In my quest to clarify the basics, I have tried to point out in a number of instances where riding is not an exact science and where there is controversy and debate. Learning to ride and learning how to train a horse are not, unfortunately, completely objective matters. There are conflicts and opposing views galore.

Ask ten people to watch a horse moving in a trot and you'll get ten different points of view. One will say take that flash noseband off — there's too much pressure on the nose. Another will comment that the horse's back legs are not moving enough, or that the horse is behind the bit.

But the fundamentals are the constants. Personally, I embrace the idea that the more you know, the more you realize that you don't know. New information, new ideas, and a deeper understanding of concepts help turn riding from a sport into an art.

As I have endeavored to present the "facts," I have certainly struggled with the thin line between what is fact and what is debatable. Sir Winston Churchill once said "the fear of being contradicted leads the writer to strip himself of almost all sense and meaning." While I'm sure some people might contradict what is being presented here, I have worked hard to describe the true fundamentals that every English rider should know. As a broad framework I hope there is sense and meaning that helps horse and human alike.

CHAPTER 1

LEARNING CONTROL

The oft-quoted line "A journey of a thousand miles begins with a single step," credited to Confucius, applies both literally and figuratively to the journey of learning to ride.

AT THE MOST BASIC LEVEL, to control a horse is to determine a) where the horse puts his feet and b) at what speed the horse moves. New riders need to recognize that they must be in control from that first step. All types of riding (e.g., Western, dressage, trail, polo, etc.), regardless of purpose, have two questions in common: Where am I going on this horse? and How fast is the horse moving?

Being in control is a thorny issue with several layers of complication. What does it take to control a horse? Brute force? Immense strength? Neither. It takes a good, balanced riding position and the ability to use that position to *communicate* with the horse. This universal ideal of communication is based on science and the nature of the horse, not on a particular method or system.

In many riding schools in Europe, a beginning rider does not immediately control the horse's forward motion and direction. Instead, she spends many lessons on the lunge line (see box on page 3), learning balance and establishing a good seat. After the seat is established, she learns proper leg position and aids, followed by hand position and aids. Lungeing allows her to gain the necessary knowledge and skills to communicate with a horse and control the ride.

In the United States, in contrast, riders usually start with the reins in hand and are instructed from the beginning how to use the hands and legs to move the horse forward, turn him, and stop him.

Be in Control from the Start

With either method, the goal should be for a rider to learn how to use a balanced, positive position to tell a horse where she'd like to be going and at what speed. Being in control allows a rider to relax and gain confidence. Unfortunately, this is often the paradox in learning to ride. Without confidence, the rider can't relax. A tense, nervous rider isn't able to use her position to communicate with her horse and, consequently, she won't be in control. This, in turn, makes the rider even more tense and nervous.

DRESSAGE MEANS TRAINING

What many people just getting involved in riding don't realize is that the word **dressage** is simply French for "the training of the horse." Yes, it is also a discipline like jumping or eventing; however, at its most simplistic level, it is the training and basis for all riding and disciplines.

Whether you know it or not, you're doing dressage whenever you ride. You're doing dressage if you're in a Western saddle or riding bareback. By working with your horse, you are, by definition, your horse's trainer. Everything you do, consciously or unconsciously, is training your horse.

#1 Control from the First Step

I T IS A COMMON phenomenon that even after a rider establishes the physical basics of position and use of aids, she is still inclined to take a back seat to a horse's decision about where they are going. It often begins the moment the rider steps up on the mounting block.

Walk into any barn and observe the number of horses who walk away from the block the second a rider swings her leg over the saddle. The horse should stand while the rider is organizing the reins and stirrups.

The horse needs to wait for the information about *where* and *when* to make that first step, but this will happen only if the rider has a clear intention of what she wants. Without forethought and planning, what follows is often a mindless meandering around the arena. The moment you get on a horse, you set the tone for your ride. For example, you should know in which direction the warm-up will begin. The rider with no plan lets the horse make the choice.

When a rider turns the ride over to the horse, it's often because, although physically able to be in control, she is not *mentally* doing the job. Riders who

let the horse make the majority of the decisions have, unfortunately, "trained" the horse to be in charge.

Have a Plan When You Ride

Every rider needs to take responsibility for that first step, and for every subsequent step, of the journey. You must have a plan in your overall training and riding journey, not only for the big picture, but for the day-to-day, one-step-at-a-time picture. You must know what you're going to do in the arena on any particular day.

In the big picture, whether you are just learning to ride or

▼ Control begins well before you mount your horse. Have a plan before you get on.

are determined to ride better, you need to lay the groundwork for your intentions. Are you interested in showing or riding for pleasure? Are you thinking that, one day, you might want to be a professional?

If you are working with a particular horse, what are your expectations for him? Do you hope to show First Level next year, or would you like to try a combined training event? Regardless of the answer, laying a good foundation and understanding the fundamentals of riding is the necessary first step.

In your daily planning, come to the barn with an idea of what you'd like to do that day. You might concentrate on cleaner transitions or more effective half halts. Perhaps you have a week-by-week schedule written out that serves as a framework for your goals.

In the actual ride, break it up, with a warm-up that is at least 10 minutes. Figure on a work session that lasts somewhere in the neighborhood of 10 to 20 minutes. Count on a cool-down period of 10 minutes. This structure will help you focus your ride time into manageable bits.

The moment you get on a horse, you set the tone for your ride. For example, you should know in which direction the warm-up will begin. The rider with no plan lets the horse make the choice.

Keep in Mind

Riders make mistakes on both sides of the time-management spectrum. At one end, there are the riders who get on and off in 20 minutes because they can't think what to do. (This, by the way, is often the consequence of an instructor who barks orders all through a lesson, and never gives the rider a chance to think for herself. When the rider tries to ride on her own, she finds that her mind draws a blank because she is so dependent on her trainer. A good instructor encourages and develops an independent rider.)

At the other end is the rider who doesn't know when to quit. She drills for 2 hours, maybe more. A horse (and often the rider) will lose focus as he becomes fatigued. An hour, give or take a few minutes, is the recommended maximum time for arena work. Hacking in the fields and trail riding are less taxing mentally and oftentimes less physically intense, so sessions can be longer.

As you work toward your goals, build a solid base by being organized and acting with intention for your journey. In your riding and education, proceed with clarity so that your horse understands your desires. You will be a confident, relaxed rider and in control of the first step and every step that follows.

THE LOWDOWN ON LUNGEING

The first question most people have is, How do you spell this word? Is it lunge or longe? Lungeing or longeing? Both are acceptable, but either without the "e" is not (despite what the spell-check on your computer says).

Lungeing benefits both the horse and rider in a number of ways. The basic method is for the horse (with or without a rider) to work at a distance from the handler, moving in a circle at the end of a lunge line that is attached to the halter or bridle. Generally, the horse is worked on a 20-meter circle. He can be worked in a walk, trot, and canter, and in both directions.

Lungeing is an ideal way to help the rider work on her own position and balance because the instructor controls the horse's speed and direction, allowing the student to focus on one or two issues at a time. Lungeing can help a timid or nervous rider gain confidence and learn to relax. It is extremely important that the horse is a well-schooled, knowledgeable lunge horse before putting a student on top of him.

Done correctly, lungeing can also be good for the horse, for many reasons. Young horses are often lunged as part of their training program. It builds trust and confidence and allows the youngster to get used to the saddle. It teaches bending, flexing, and correct contact. Putting a horse on a lunge line before a riding session warms him up and loosens his body. Many people lunge a "hot" horse to take a bit of extra energy off before mounting.

CHAPTER 2

RIDING POSITION

"Riding is the art of keeping the horse between you and the ground."
I'm not sure who coined that phrase, but I do know that Sir Isaac
Newton was onto something with his Law of Gravity, because every
time you get on a horse, you challenge it!

I CAN ONLY IMAGINE that the first humans to conjure the notion of actually getting on a horse must have instinctively wanted to lie horizontally on the horse's back. You see this in new riders who throw themselves forward, hang onto the neck, and become as horizontal as possible in the hopes of preserving life and limb when there is trouble in the balance department. After all, who in their right mind would want to mix the horizontal surface of the horse's back with the vertical position of the human body?

The laws of physics argue against this concept — at least until you gain more understanding about the physical relationship between horse and rider. Riding a horse is teaching your brain to let go of the self-preservation mode (i.e., lie flat and hang on for dear life) and use the conscious, thoughtful brain to say, "I can stay sitting up and in balance on top of this creature even as we fly across the ground and through the air at remarkable speeds."

The Horse Stance: A Hint from Tai Chi

When you're in the saddle, you should be able to drawn a straight line that connects your head, hip, and heels. This balanced riding position is also a martial arts position, dating back to the earliest development of Tai Chi. It's actually called the "horse stance" and it's the most common of all the Tai Chi and Qigong postures. In Tai Chi, the horse stance is used to transition from one exercise to another and it helps to maintain proper flow of chi (energy).

When you are riding in this balanced position, you would land on your feet, in that same position, if the horse suddenly disappeared from beneath you. The same idea pertains to a rider in a jump position. With the body balanced over the hips and feet, the rider would wind up standing on the ground if her horse evaporated.

The Other Elements of Position

The subject of correct position also includes key insights into the form and function of the seat, legs, and hands. In the development of a rider's position, the seat is the top priority, with the leg coming in second, and the hand third.

◀ The Horse Stance, a fundamental pose used in Tai Chi, is the same balanced position used for riding.

#2 Dressage Seat

THERE'S A SAYING in Germany that only a rider who *sits* correctly can *ride* correctly. (You can always count on the Germans to be direct and to the point.) This statement reflects a key idea in learning to ride: form really does lead to function. A correct seat is a nonnegotiable requirement of a rider's education, and there are no shortcuts to developing it.

The incomparable Anne Gribbons (United States Equestrian Federation [USEF] dressage technical advisor; Fédération Equestre Internationale [FEI] 'O' Judge; and Pan Am silver medalist) once wrote something along the lines that every rider should be on the lunge line until the horse, the rider, and the instructor are bored to tears.

Gribbons, who trained in Europe, knows that every rider must dedicate considerable effort to her seat, as it is the foundation on which the rest of the riding education builds. A rider without a good seat is like a tower built on loose sand — there's no stable base on which to build (in this case, a strong, balanced rider). In European schools, it's considered bad form to have a novice rider take on the responsibility of using her hands correctly if her seat is not yet developed. Riders start on the lunge line and do not pick up the reins until the seat is well established, whether it takes 10 lessons or 110. In other cases, particularly with children, the rider does hold the reins, but the horse goes around the school on side reins so that the rider can concentrate on her seat, with no obligation to have anything to do with the correct contact (see page 95) and frame of the horse.

What Is It?

The "dressage seat" is a balanced way of sitting atop a horse that lines up the head, hips, and heels in a vertical line. Also called an "independent" seat, it is described as soft, effective, and elastic. The rider's seat bones and pubic bone form a triangle in the saddle, with the weight of the body equally distributed over the two seat bones. When a rider is in balance and her seat moves freely with the horse's motion, three important things happen:

▶ The rider is not blocking or inhibiting the horse's movement.

▶ The rider can intentionally influence the horse's movement with her seat.

▶ The rider can effectively control her legs and hands.

On the spectrum between being loose (allowing the horse to go forward) and tight (inhibiting forward motion), a rider working on her seat often suffers from the extremes. Finding the right mix between firm and loose is like trying to mix hot water with cold to come up with just the right temperature.

In broad terms, a horse will react to looseness by moving more forward and to tightness by slowing down. This is the essence of points one and two above.

SEAT FUNDAMENTALS

As with most things that are difficult to master, there is more to a correct independent seat than meets the eye. If you lined up the world's best riders, you would see that everyone has her own version of a good seat, and that there is no one "correct" seat. There are, however, several fundamentals that every successful rider shares and that allow them to develop a highly functioning seat:

- balanced and coordinated muscles

- relaxation

- confidence

- sensitivity to the horse

On one end of the spectrum is the rider who sits like a blob with no muscle support. On the other end is the rider who is as rigid as a 2×4 board. Somewhere in the middle is a place of positive tension where you are soft, but firm. When a rider does find the right ratio, the horse's motion owns the seat, meaning that the rider's seat melds with whatever motion flows up through the horse and doesn't fight it or block it. At the same time, the rider has the freedom to use her hands and legs independently from the motion of the seat.

▼ In a correct balanced seat, the rider's head, hips, and heels are lined up.

Common Seat Errors

PROBLEMS IN SEAT POSITION are universal. They are so common that they have their own names: chair, fork, collapsed hip, hollow back, and round back. As a rider searches for her seat, it's normal to see her dealing with one problem only to develop another. Sometimes riders struggle with a combination of these issues.

In both chair- and fork-seat positions, for example, the rider finds balance a problem. When balance is compromised, a rider struggles to use her weight and leg aids effectively, because the seat and legs are not where they need to be to give correct aids. Making matters worse, her hands can't act independently from the rest of her body because they're being used in an effort to find balance.

Chair Seat

In the chair-seat position, the rider looks as though she's sitting in a chair. Her legs are out in front and the seat is pushed back at the deepest point of the saddle, toward the cantle. Her thighs and knees tend to be drawn up and are tight. This often occurs if the stirrups are too short or the rider isn't using the knee as a flexible joint.

Fork Seat

In the fork-seat or stiff-seat position, the most notable problem is that the lower leg has slipped back and is not under the rider. As the leg slides backward, the rider's shoulders pitch too far forward. This is the effect of the body trying to counterbalance itself. Riders whose stirrups are too long will fall into this position.

▼ CHAIR SEAT

▼ FORK SEAT OR STIFF SEAT

Collapsed Hip

For many reasons, among them curvature of the spine, strong side versus weak side, and bad saddle fit, riders can easily fall into the trap of sitting crooked in the saddle. This problem blocks a horse from moving freely and correctly, and inhibits a rider from being able to use correct aids.

For example, if a rider collapses her hip onto the right seat bone and always pushes a horse with more weight on the right side, a left-lead canter (which requires a shift of weight onto the left seat bone) could prove to be difficult.

It's quite common to hear a rider say, "Oh, my horse always has trouble with his (blank) lead," when it's quite possible that the rider is to blame because she's sitting to one side and not letting her horse canter on that lead.

Over time, saddles (and horses) can become twisted and collapsed to one side from a rider who is unaware that she is perpetually sitting with a collapsed hip.

Round Back

Too much time spent hunching over computers and slouched in chairs leaves many riders with weak pectoral and abdominal muscles and a constant slouch. Slumping shoulders and a rounded spine transfer directly to poor posture in the saddle and put the rider in a weak position. A rounded back means a rider is able to convey information to the horse through only the arms, instead of accessing and using her core muscles.

A rider's core refers to the broad group of muscles that run the length of the trunk and torso. When a rider engages these muscles, they stabilize the spine, pelvis, and shoulder girdle. Integrating activities such as yoga, Tai Chi, and Pilates into your exercise and fitness routine is a good way of building a strong core.

Hollow Back

People wonder why, in day-to-day living, they have sore backs, and often it's the forward tilting of the pelvis that is to blame. Take that same position and put it in a saddle and you've got a person who is "perched" with the seat bones lifted from the saddle. This locked-forward position also leads to pinching with the inner thighs and even pinching with the knee. The result is that the rider is unable to move with the motion of the horse. This equivalent of a "human clothes pin" tells the horse not to move forward.

▼ COLLAPSED HIP

▼ ROUND BACK

▼ HOLLOW BACK

Improving Your Seat

THE BEST FIRST ADVICE for establishing a solid foundation is to make sure you're getting good instruction. What is the instructor's background, education, reputation? Your instructor needs to be knowledgeable and able to communicate well.

There's nothing worse than going through the trouble to learn to do something and then finding out that you've learned the wrong way. Breaking bad habits is considerably harder than establishing good ones (e.g., building a good seat) from the get-go. If you've never ridden, or you've been an infrequent rider, and get good instruction, you have the advantage of beginning correctly.

Check Saddle Fit

As you learn, or suffer through the arduous task of relearning, there are several areas of concern to address before you even get in the saddle. First, take a look at the saddle you're using. Is it appropriate for you and your horse? I put saddles into three categories: helping, neutral, or hindering. It's surprising how many people ride in saddles that get in the way of developing a good seat, and actually cause many of the difficulties.

Take the time to figure out whether the saddle fits you *and* if it fits the horse you are riding. Many riders operate under the theory that if the saddle fits the horse, that's good enough. Not so.

In the end, if the saddle doesn't fit you, you will be hurting your horse because you'll be sitting so poorly. It is necessary that you find a saddle that works for both you and your horse.

Saddle fitting is a science. Search out knowledgeable help. The people at your local tack shop can provide a good start. Ask your trainer for some insight. The biggest areas of concern are clearance for the spine and freedom at the shoulders and withers. The horse must have complete freedom of movement.

If the saddle appears to be appropriate for your horse, try to put in some ride time in it before you put your money down. This will confirm that your horse feels

▼ One tried-and-true tool for addressing seat issues is to get on the lunge line.

good in it and give you a chance to see whether you are comfortable and all the elements work for your body.

Watch Good Riders

One of the best ways to learn is to watch riders who know what they're doing, and try never to watch a bad rider. If you're a visual learner, it's very easy to copy what you watch. It's wonderful to learn from observing at clinics, and seeing people ride at shows can be a great learning opportunity. It's a truism, however, that you will start to ride similarly to the people you watch, so be careful who you are watching.

On the other side of the coin, if your trainer is a wonderful rider with a beautiful seat, watch to your heart's content. I once went to a show where a trainer was riding along with a handful of his students. I realized I could pick out his students just by watching, as they all rode just like he did: beautifully. His barn walked away with every big award of the show.

Back to Basics

As pointed out earlier, there is an extra burden on the person who already rides but has seat problems. The job of relearning is harder because of muscle memory. Your body thinks it's

doing everything right because that's what you've taught it to do, but now you need to actively ask your brain to intervene. This can be very frustrating, and you need to have patience with the process.

A lunge lesson allows you to focus exclusively on your seat because it's not your job to steer your horse and keep your horse moving. A good instructor will give you a multitude of exercises to work through.

Try Shorter Stirrups

A helpful exercise that you can do on your own is to shorten your stirrups by five or seven (or maybe even more) holes, and ride in a jumping position, even if it's

▼ Shorter stirrups can help you find your balance and strengthen your seat.

in your dressage saddle. This exercise helps you find and loosen your three important joints: hips, knees, and ankles. They each should be soft and springy, and you should work to find your balance on the short stirrups as you feel like a jockey atop your horse.

Watch Where You're Going

A sure way to improve your seat is with your eyes, as strange as that sounds. Like a flowchart, the eyes lead the head, the head leads the body, and the body leads the horse. The ideal rider is looking softly ahead, much as you do when you're driving a car. (This is opposed to looking down at the road right in front of the hood of your car.) The rider's neck is tension free, and motion flows freely up and down through the head, neck, spine, and seat.

SOFT EYES. Let's start with what happens when a rider locks her eyes on a specific point. Riders with a poor seat will often lock their necks and stare down at some unknown spot on the horse's neck or on the ground. This "eye lock" means that the rider is physically unable to flow with the motion through her seat.

The late Sally Swift, of Centered Riding fame, used the phrase "soft eyes," meaning when a rider is looking softly ahead, the head position corrects itself and the rider becomes more in balance and free. When a rider is stiff and tense in the neck and head, the entire body is unable to relax, the balance of the rider becomes questionable, and the result will automatically interfere with the horse's balance.

STEADY HEAD. Your head weighs roughly 13 pounds (6 kg). A head that is locked and off-kilter by 13 pounds (6 kg) might not seem like much, but for a horse who's trying to maintain balance under a rider, this weight shift can be serious.

As a demonstration of this, I once was handed a 13-pound bowling ball on a stick, symbolizing a head on a neck. I was amazed at the difficulty I had balancing the ball. After that, I became much more sympathetic to the horses I rode with regard to how I held my head.

Find a Way to Relax

Diagnosing the difficulties on the outside of the head is easy compared with understanding what's going on inside the head. The inner workings of the brain are perhaps the most important element because a good seat is a reflection of a rider who has a balanced body *and* mind. Mental relaxation leads to physical relaxation.

The person who comes into the arena with a negative attitude; is thinking about all the bad things that happened that day or even a week or a month ago; is riding with grinding, clenched teeth; and is worried about all the "what ifs" is not going to be nearly as successful as the smiling, confident rider who comes to her session with a positive attitude.

THE "INVENTION" OF THE CLASSICAL DRESSAGE SEAT

The history of classical riding doesn't follow a straight path but rather a winding road with many influences over the centuries. Everyone agrees that it was started by the Greek writer, Xenophon, in 400 BCE. Since that time, various masters have made their mark on the art of riding.

Among them, an eighteenth-century master is credited with inventing the dressage seat. François Robichon de la Guérinière (1688–1751) was a French riding master who wrote a book published in 1733 that states that the rider must have a good seat to have a soft, light hand. (Sound familiar?) To this day, much of the everyday training of the famed Spanish Riding School is based on this book.

Make Your Seat a Priority

A final thought on a good seat. Many people think that once you have a good seat, that's it — you've got it forever. Unfortunately, that's not the case. I know of one professional at the highest level who focuses one ride a week solely on her seat and position. This is because a good seat needs maintenance, like a house or a car. If you don't fix the little problems and have regularly scheduled tune-ups, it will fall apart.

Before we leave dressage seat, you may be wondering, "What about other positions," such as the light seat used for jumping? Yes, that's a fundamental of good riding too, but before you can master the light seat, you need to firmly establish the dressage seat, legs, and hands as a foundation.

There's nothing worse than going through the trouble to learn to do something and then find out that you've learned the wrong way. Breaking bad habits is considerably harder than establishing good ones from the get-go.

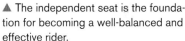
▲ The independent seat is the foundation for becoming a well-balanced and effective rider.

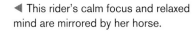
◄ This rider's calm focus and relaxed mind are mirrored by her horse.

#3 Leg Position

Once a rider's seat is established, the focus moves to the form and function of the legs. Not that one's seat can be fully separated from the leg, but there are specifics to the leg position that need to be addressed.

Before we talk about the leg, a word about terminology: a "leg" in riding means something slightly different from that in most other contexts. For most riding instructors, the leg means only the lower portion of the limb, from the knee down. The thigh is part of the seat.

The Ideal Leg

Remember the old Western movies with the bowlegged cowboy walking down the street? Although we don't want to create riders with deformed legs from too much time in the saddle, a person who rides well is not

◀ A good analogy for how the legs should drape would be spaghetti that's cooked *al dente*: soft, but just on the edge of firmness. The wrong extremes would be uncooked spaghetti (stiff as a board) and overcooked spaghetti (wet noodle). When the leg is correct and hanging softly, the knee is bent at an angle that allows the leg to move forward and back and up and down.

actually sitting on the horse as much as draping their seat and legs around the horse. The shape of the seat and legs should remind you of a wishbone.

THE HIP JOINT needs to be open to allow a soft curving of the flattest part of the thighs down to the knees, then a lower leg that encloses the horse with your body.

YOUR TOES AND KNEES should be pointing mostly forward, with an ever-so-slight turn to the outside.

THE HEEL needs to be springy and is lower than the toe. How much lower? That depends. If you are in a jumping position (see light seat on page 25) the heel tends to be deeper to accommodate a different need for balance. In dressage work, the answer depends on how much motion the horse is giving to the rider.

If the horse is in a buoyant, energized movement, the rider's heel should be moving up and down to match the rhythm of that gait. The movement in the ankle is a good sign that the rider is using the ankle joint. That joint, along with the knee and hip, is the equivalent of a rider's shock absorbers.

THE FOOT should be in the stirrup to the ball of the foot (for jumping, move it a bit deeper) and resting fairly evenly and lightly in the stirrup irons.

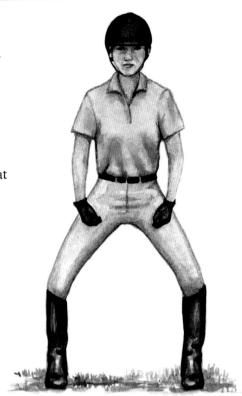

PROPER STIRRUP LENGTH

How long should the stirrups be? That's a matter of experience, intent, and comfort.

If you're riding in a dressage saddle, a good rule of thumb is that the stirrup iron should fall just a hair below the ankle bone. In an all-purpose saddle, aim for the stirrup iron hitting right about at the ankle bone.

If you are a less-experienced rider, a notch above the ankle bone is good because a shorter stirrup is always a more secure stirrup.

For those who are riding in a close-contact (jumping) saddle, to accommodate the forward nature of the saddle, the stirrup might need to be 2 or 3 inches (5–7.5 cm) above the ankle bone.

Generally speaking, riders who are struggling with a lack of spring in the joints (hips, knees, ankles) should school in stirrups that are shorter, rather than longer. After a better leg is established, you can always take the stirrups down a hole or two. If you're debating just what is right for you, the best answer is usually to go shorter.

This guidance might not mesh with what is actually being practiced in your barn. I see far more dressage riders with stirrups that are too long rather than too short. When a rider is posting off their toes and reaching for their stirrups, it's obvious that the stirrups need to be shortened. In just about every case when I make positional adjustments with a student, the first thing I do is shorten her stirrups.

Common Leg-Position Errors

In the early stages of riding, people assume that their feet in the stirrups and their hands on the reins will be their savior in any unbalanced moment. It's not uncommon for novice riders to take a death grip on their stirrups (to the point of trying to curl their toes inside their boots to hang on) and a death grip on their reins.

But a balanced position, with a soft, independent seat, is the key to a solid riding foundation. If you get into trouble on horseback, your seat and leg, not your stirrups or your hands, will be your safety line. Many of the common problems with leg position are really symptoms of a seat that needs more work.

It bears repeating that a balanced position depends on the three main joints: hips, knees, and ankles. The problems in leg position often boil down to a stiffness in any one (or a combination) of these joints.

Pinching Knees

Many problems with leg position start at the knee. When a rider uses a tight knee joint, it results in inward pressure on the saddle. The consequence is something akin to using a Vise-Grip on the horse. The pinching knee cuts off the effectiveness of the lower leg and blocks the horse from

moving freely. (It also leaves the rider's legs very tired very quickly.)

Open Knees

The opposite issue is the knee that falls open and away from the saddle. Sometimes you can spot this problem as the rider is walking into the barn, long before she gets on the horse. The person who walks on the outsides of her feet will often ride with the weight of her foot resting on the outside edge of the stirrup. Getting an effective leg on the horse is nearly impossible from this position.

▼ PINCHING KNEES

▼ OPEN KNEES

Stiff Ankles

A stiff ankle joint and tight calf muscles are most likely the cause of the rider balancing on the front of her foot. This problem is also widely seen in the rider whose stirrups are too long. When a person is riding off the ball of her foot, she tends to be catching her balance with the knees or, worse yet, using her hands for balance.

Foot Position

Pinching with the knees often leads to the foot being too far into the stirrup iron. When the rider exerts pressure on the saddle with her knees, she is unable to stretch her lower leg. This makes her off balance and she won't be able to use her leg on the horse. This problem becomes a bigger issue at the canter. In an effort to stay on the horse, she grips at the knee, her heels come up, and her foot slides through the stirrup.

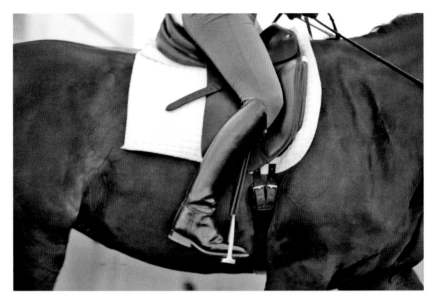

▲ STIFF ANKLES

▲ FOOT POSITION

Solutions to Leg Problems

The best solution for most leg problems is to find your way back to the lunge line. A good instructor will have you work on the independence of your seat, and along the way add more softness to your leg.

Ride without Stirrups

Whether on the lunge or not, you can practice taking your feet out of your stirrups to see how long you can maintain correct leg position. It will be hard at first, but building up this ability is a key element of a strong foundation. Walking, sitting trot, posting trot, and even cantering without stirrups will make a huge difference in how you position and use your leg.

Open Your Hip Joint

Another helpful exercise is a simple hip release. Of the three important riding joints — hips, knees, and ankles — the knees and ankles are hinged joints and the hip is a ball-and-socket joint. The hip is intended to move all around, but how often in any given day do you actually use the hip as created by nature?

Picture yourself sitting in a car seat, at your desk, or even taking a walk. Most often, we treat the hip like a hinge joint, pivoting front to back, but not side to side.

To sit well in a saddle and use your legs properly, you must open the hip joint to allow it to function as nature intended. Have someone on the ground gently lift your leg to release tightness in the joint, or do it yourself by lifting both legs up in front of you and putting your knees together in front of your saddle.

As you put your legs back into a position, take them out to the side first and then down. In all likelihood, you will find that your stirrups now feel a little too short. That's a sign that you've released your hip joint, dropped your seat bones more into the saddle, and achieved more of the desired wishbone shape.

▲ Riding without stirrups is an excellent way to develop the correct position for your leg.

▲ A word of caution: some people find this exercise painful and, as with all exercises, you must use good judgment and common sense. If you find yourself experiencing anything more than the slightest discomfort, don't do it.

Test Your Leg

Riding with your feet out of the stirrups is a great way to test for correct leg position. What happens next tells a lot about what the leg is really doing. Can you maintain a riding position or does your leg collapse and hang like a sausage?

If your toes drop down when you come out of the stirrups, your knees lose their bend, and your legs just hang, you are using the stirrups for control of leg position. Think of your stirrups as little more than a placeholder for the foot. Your leg position should be the same with or without stirrups.

▲ I suspect the reason that riders tend to set their stirrups too long is our vision of world-class riders with long, beautiful legs. We think, "Long legs equal great dressage riding." But look carefully and you'll notice that these riders still have a springy knee and ankle, so their stirrups aren't as long as they seem. It's just that the leg position is correct, and therefore, the rider looks as though she's using a long leg.

STIRRUP IRONS

This won't make me any friends in the tack manufacturing and sales industry, but think twice before you walk into the tack store and drop a bundle on some newfangled stirrup irons. Irons that are offset, hinged, angled, twisted, turning, bouncing, flexing, or any combination of these will not fix your leg problems.

Some can mask the true nature of your leg issues, which is probably your seat. Depending on the configuration, they can mechanically set your foot or the angle of the ankle bend, or provide a spring that should come from your joint. Spend your money wisely, perhaps on more lessons, and especially lunge lessons, that truly address the issues.

There's a possible exception for people with joint injuries and other limitations such as arthritis. For some of these riders, specialty stirrups might help.

#4 Hand Position

▲ Good hand position extends in a straight line from the elbow through the hand to the bit.

▲ Notice the softly closed fingers and the rein held snug between the fourth and fifth fingers.

THE HAND IS the final piece of the puzzle for a correct riding position. The reins form our connection to the horse's mouth and must be under the rider's control, not at the mercy of the horse's motion. The importance of how we hold our hands and what we do with them can't be overstated.

What Is Good Hand Position?

Your hands should hold the reins in an upright position, with your thumbs as the highest point, creating a straight line from the horse's mouth to your hand to your elbow. The reins should be held in your hand between the fourth and fifth fingers, and should be positioned close to the webbing of the fingers, with the fingers closed. Your thumbs form a lock on the reins as you press the rein between the thumb and the index finger.

What's the reason for this position? It's the most direct, uninterrupted line for communicating with the horse through the reins, and for keeping that connection consistent. Until you firmly establish your hand position, you'll be struggling to convey the messages you'd like to send to your horse. (See page 39 for more on using the rein aids.)

Common Hand Errors

Anyone who thinks riders aren't creative hasn't seen all the different ways we have of making mistakes with our hands. We have lots of different areas to worry about: fingers, wrists, elbows, shoulders, and seemingly everywhere in between.

Open Fingers

Open fingers are probably the single most common mistake. I've gone as far as to consider installing a billboard-sized neon sign in our arena that says, "Close your fingers!" Without closed fingers, a rider can't communicate with a horse in any consistent manner. Think of it as communicating with your horse by cell phone while driving through a rural area. When you let your fingers fall open on the reins, it's as though "the call" got dropped.

How much pressure should you use in your hand? The baby bird analogy is used by countless instructors to convey the idea of what the hand should feel like. Pretend you're holding a baby bird in each hand. You need to keep the fingers closed enough so the baby bird doesn't fly away, but if you make a white-knuckled fist, you're going to squeeze the living daylights out of the birds. Don't kill your baby birds, but don't let them fly away either.

THE SOLUTION. To fix the open-finger syndrome, you either have to actively engage your brain to think about keeping the fingers closed or be frequently reminded by an instructor until it's firmly ingrained. One solution that puts a rider on the right road is for her to carry small stones inside her closed hand (along with the reins). She will feel the stones (a reminder) and will need to keep her fingers closed in order not to drop them. If she keeps dropping the pebbles, it becomes a significant reminder of just how often her fingers are falling open. If she loses the stones from only one hand, it indicates that one hand is weaker than the other.

Piano Hands

Piano hands, a classic rein-holding error, refers to hands that face down as though the rider were playing the piano. In this position, the elbows automatically move out from the ribcage and the connection to the horse's mouth becomes insensitive and stiff. Piano hands make the rider incorrectly use the whole arm or

▼ OPEN FINGERS

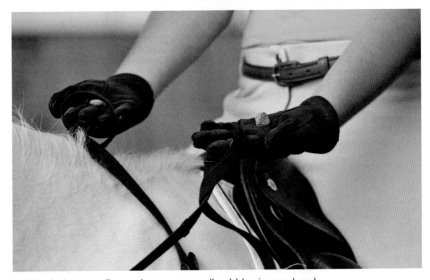

▲ One fix for open fingers is to carry small pebbles in your hands.

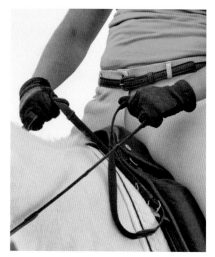

▲ PIANO HANDS

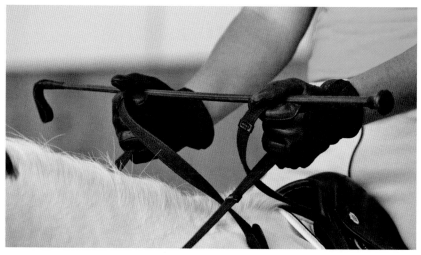

▲ Try fixing piano hands by imagining that you are holding a tray full of drinks.

the body to give aids. Having your hands in the piano position tightens up the forearm, elbow, and even shoulder.

Correct aids should be a sensitive use of the hand and wrist, plus a good connection to an elastic elbow, all of which come from softness from the wrist through to the shoulder. Now you know why instructors are often heard saying, "Thumbs up."

THE SOLUTION. Imagining carrying a martini tray is an excellent way to reinforce correct

alignment. With your hands properly positioned, have someone slip a crop or slender whip under your thumbs. Visualize the crop as a tray that you can't tip in any direction or everything will fall off. This exercise helps establish the correct muscle memory for keeping the elbows softly at the ribcage and keeping the hands level with thumbs on top.

It will be harder to steer during this exercise (until you have experience with the correct position), and you should avoid more complicated work.

Stiff or Broken Wrists

Although they sound like opposites, both "stiff wrists" and "broken wrists" refer to a rider without mobility at the wrist, as if she were wearing a cast. When the wrist is cocked or locked into a position, there will always be contact issues because the connection from the horse's mouth through the reins will end in the wrist, preventing communication from extending through the wrist into the elbow and up to the shoulder.

▼ STIFF OR BROKEN WRISTS

▼ A visual reminder can help you properly straighten and align your wrists.

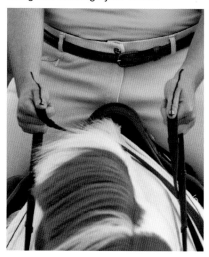

THE SOLUTION. Fixing stiff or broken wrists requires some creativity. One exercise that I've had some luck with is to have a rider take off her gloves. Using a washable marker, I draw a line down her wrist and along the top of her hand and thumb, and put an arrow on the thumbnail. This arrow needs to follow the line from the rein to the horse's mouth. The rider with problematic wrists soon recognizes, on her own, how often the arrow is pointing in some direction other than that of the horse's mouth.

Locked Elbows

A locked hand position quite often accompanies a locked elbow position. The rider who does not have an elastic elbow joint is simply not moving with the horse's motion and has the unfortunate consequence of bumping the horse in the mouth. Many times, the rider isn't even aware of how much she is freezing her hands and elbows, as if she were a statue.

A good exercise is to have someone physically hold the elbow joint and make it follow the connection of the rein as it's moving. The locked elbow then releases and something clicks in the rider's brain.

This exercise may need to be repeated a couple of times in a lesson, and over the course of several lessons, but it's imperative that the rider starts to *feel* what an elastic connection through the elbow is all about.

Check Your Thumbs

When working to develop a soft, elastic connection, a remarkably little fix — a simple thumb adjustment — can result in a huge payoff. When the thumbs are closed down flat on the top of the rein, the arms tend to be tight. When there is a tiny bend in the thumb, with the nail of the thumb pointing just slightly down into the

▼ LOCKED ELBOWS

▼ Try this exercise to soften rigid elbows.

rein, the whole arm softens and becomes lighter. This tiny change, which translates down the rein to the horse's mouth, can have a huge impact on your horse's acceptance of the contact and his sensitivity to the aids.

Don't Cross the Neck

Among the many bad habits with the hands, crossing the rein over the horse's neck is one of the worst. When this happens, the action on the bit and, therefore, on the horse's mouth, gives mixed and confusing information.

The horse's neck has no choice but to constrict, which makes him tilt his head at the poll. To address this habit, visualize a glass wall running along the horse's mane. The hands might move as far as the mane, but they can't go through the glass wall.

▲ DON'T CROSS THE NECK

ADJUSTING THE REINS

When you're attempting to keep your horse softly connected through the reins, start with addressing your rein length. Shortening your reins by opening your hand and grabbing the rein closer to the horse's mouth disrupts the contact that you've been working to keep steady. It can also send the message to your horse to zoom forward, since you've just thrown the front door wide open.

The correct way to change rein length while riding is something I call the pinch and slide method. If both reins are too long, you pinch one rein with the opposite hand, slide your hand down to the point that seems like the correct length, and then repeat on the other side. If only one rein needs to be shortened, then you just pinch that rein and slide down.

If the reins are too short, slightly open your fingers to allow some rein to slide through, and then close your fingers again to secure the connection.

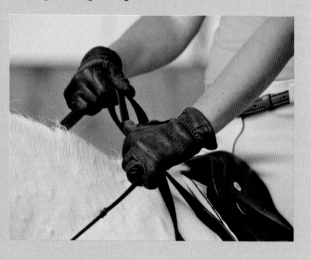

#5 Light Seat

MUCH TO THE CONFUSION of students everywhere, different instructors and different books have subtly different definitions and expectations for the following terms: light seat, half-seat, forward seat, two-point position, jump seat, and gallop position. Other instructors and books use them as though they mean the same thing.

For the purposes of this book, and for a broad understanding of the concept, "light seat" is probably the best term, since it best suggests the riding position we're actually trying to have: a position that takes some of the weight from your seat and transfers it to your legs. All the other terms are just variations on the same basic idea.

What Is It?

A light seat involves taking your seat out of the saddle and putting weight on your stirrups and especially your thighs, thus becoming lighter in the saddle. More leg means less seat. Most

people will identify a light seat with the world of jumping, and indeed that's where the position originated. An Italian equestrian by the name of Federico Caprilli (1868–1907) revolutionized the jumping position when he determined that the light seat

was the best technique for jumping a horse. It gives the horse the freedom he needs to get over an obstacle most effectively, and it provides the rider the balance and support she needs to be secure while staying out of the horse's way.

▶ A light seat can be used even in a dressage saddle, to release the horse from the weight of the rider's seat.

When to Use Light Seat

It's obvious that the light seat is the correct position for jumping. But the light seat also has important uses even if there's not a jump in sight. Trainers frequently request the light seat when looking to enhance a rider's basic position, balance, and skill. They also employ it to build a horse athletically and gymnastically.

In most training and schooling situations, you don't have to be in a jumping saddle to utilize the light seat. The stirrups on your dressage saddle should always be of a length such that you could get up in a moderate light seat without any problem.

Benefits to the Rider

For the rider, there are many benefits from working in the light seat. This position uses different muscle groups and asks for a broader sense of balance on the horse's back. As pointed out in the discussion of dressage position, the ankles, knees, and hips are a rider's shock absorbers. Unless you use these joints elastically, you will struggle with a laundry list of problems, including loss of balance and ineffective aids.

The light seat allows the rider to find a deeper heel, springy ankle joint, mobility in the knee, and a hip joint that is open and supple. It also develops balance over the feet. Making the transition between a rising trot and a light seat and then back to a rising trot again challenges and builds a rider's balance and coordination.

Benefits to the Horse

For the young horse, a moderate light seat (meaning just making the seat lighter in the saddle by taking a bit more weight onto the legs and upper thighs and actually coming up from the seat of the saddle) can be used when riding to give him the freedom to move and find his natural rhythm. Stiff horses and older horses can be warmed up in this version of the light seat as well. It helps to give their backs time to get moving before having to handle the full weight of a rider.

The moderate version of the light seat can also be used for working over ground poles, cavalletti, when doing hill work, and when hacking in the fields. Cross-training is built on the idea that every horse is healthier in mind and body if he is schooled in a variety of ways and settings. Working in and out of the light seat opens the door to these possibilities.

▼ A moderate light seat is used when you just need to get off your horse's back a bit.

▼ In the more pronounced version, the rider comes fully out of the saddle and the hip, knee, and ankle joints close and open with the motion of the horse. This version is used for grid work, jumping courses, and when galloping and jumping cross-country.

Correct Light Seat Position

Sometimes riders will try to tighten the muscles in the leg, pinch at the knee, or brace on their feet to hold themselves up out of the saddle. Any of these mistakes will interfere with balance. As with the dressage seat, the light seat should be at the balance point, such that if the horse should suddenly evaporate the rider would be standing on the ground and not falling face-first in the dirt or backward onto her rear. To get the right idea of how to correctly ride in the light seat position, look at each joint individually.

THE HEEL is generally going to be lower in a light seat than in a dressage seat. Imagine stepping down into your heels, rather than drawing up the toes or jamming the heels down. Your foot can be a little farther into the stirrup than that in a dressage seat, but still with equally distributed weight across the stirrup pad.

THE KNEE should not be pinching in toward the saddle, but closed firmly against it. It's the difference between holding a grape between your fingers and squeezing the grape so much that the juice and seeds squirt out.

THE HIP should be bent at an angle that roughly matches the angle at which the knee is bent. These two joints are intrinsically dependent on each other, so that if the hip angle opens, the knee angle opens to the same degree, and vice versa.

THE BACK is flat and the eyes are looking forward. Your position is always affected by where you are looking, so don't focus at the ground or at your horse.

THE REINS still form a straight line of contact from bit to hand to elbow, but will be somewhat shorter to maintain the straight line.

Although dressage saddles will accommodate the moderate light seat, getting a deeper, more extreme light seat in a dressage saddle that has large knee or thigh blocks is a bit difficult. It's easier to use either an all-purpose or close-contact (jumping) saddle; both will provide ample room for your leg as it contracts and pushes forward at the knee.

▼ A good light seat allows this rider to balance over her horse and move with him through the jump.

Common Light Seat Errors

▲ RIDING BEHIND THE MOTION

▲ RIDING AHEAD OF THE MOTION

To be in a good light seat, a rider needs to use the hip by actively bending at the joint. And the rider has to do it regularly, until it becomes automatic. When the hip joint is bending, the rider can fold into a light seat with ease and maintain the balance point on the horse. The most common difficulties experienced when riding in a light seat are either falling behind the motion of the horse or getting ahead of it.

Riding Behind the Motion

When the rider falls behind the motion, the lower leg is too far forward, the seat is too heavy in the saddle, and she tends to pull back on the horse's mouth. A good way to check yourself is to briefly look down. In a correct light seat position, you should see only the very tip of your boot ahead of your knee. If you see your toe or more, it's quite likely that your stirrups are too long and you are riding behind the motion of the horse.

Riding Ahead of the Motion

When the rider is too far ahead of the movement, the lower leg is slipping back, the heels are coming up, and the seat is too far to the front of the saddle or even beyond it. Check to see whether you are actually over the pommel (front) of the saddle, which is what you want to avoid.

You want your seat to be above the actual seat of the saddle. If you're too far forward, the problem points to stirrups that may be either too short or too long. Too short pushes the rider high above the horse and the lower leg slips back as a counterbalance. Too long can also cause the leg to swing back, with the rider falling on the neck of the horse.

Cat Back

One problem is the rider's bending at the waist but not using the hip joint. The rounded or roached back that results is often descriptively called "cat back."

To fix this problem, try this exercise at the halt (or even standing on the ground): bend and engage your hip joint until you feel a pull on your hamstrings and a stretch in your quads. This means your hip is bending.

Duck Butt

The opposite problem is the rider's having engaged the hip joint to the point of locking the pelvis forward, causing the back to arch and become stiff and hollow. This is often referred to as "duck butt" because of the extreme position of the rider's rear end.

To fix, try relaxing the small of your back and not stressing your hip joint.

Solutions to Light Seat Problems

Finding the way to a light seat that is well-balanced starts with the rider practicing going into the light seat and then sitting back down while the horse stands still. Without the horse moving, you get accustomed to the feeling of moving up out of the saddle and back down.

This is balance *without* motion. As your comfort zone builds, you can try your balance *with* motion. First try the light seat in a walk, then move up to a trot, and eventually try to canter in light seat.

▼ CAT BACK

▼ DUCK BUTT

CHAPTER 3

THE RIDING AIDS

"Aids" can be a confusing term, particularly for beginning students. The aids are the primary way we communicate with the horse and ask him to do things that we want him to do. As a subtext, this communication is helping or "aiding" the horse to move in a better way.

RIDERS HAVE three main or primary aids: seat (or weight), legs, and hands. The auxiliary or secondary aids that we can use to back up our primary aids are the whip (or crop), voice, and spurs. It is paramount to understand the order of the aids. Weight aids come first, then legs, then hands, followed, if necessary, by any combination of whip, spurs, or voice.

Using aids correctly involves variations in intensity, placement, and degrees of intricate subtlety. I like to use the analogy of a symphony orchestra, with each aid being an instrument in the orchestra. Every member of the orchestra must play the same music and be on the same page, but each has its own score and time to play in concert with the others. Sometimes it is time for an aid has a time to play a solo, with the orchestra as a backdrop. There are moments for *forte* (loud) and *pianissimo* (soft).

The beauty of the music happens when each musician works together to bring the music to life in an artistic display of harmony. Correct aids are the same, with the rider's brain leading all of them as the ultimate great conductor. Learning "to play music" with your aids is perhaps one of the most profound reasons that so many people find riding such a worthwhile endeavor. It's hard, but when it's right, it's amazing.

The goal of the beautiful orchestration of aids is addressed by the FEI rulebook in the opening section on dressage: "The object of dressage is the development of the horse into a happy athlete through harmonious education. The horse gives the impression of doing, of its own accord, what is required.... [T]he horse obeys willingly and without hesitation and responds to the various aids calmly and with precision, displaying a natural and harmonious balance both physically and mentally."

This is true for all forms of riding, not just dressage. Contrary to what some people think, the aids are the same for all types of riding because they're based on the nature of the horse. It's simple science: biology, physics, and the unbreakable laws that govern nature govern the aids, too. Just as dogs like being petted whether they're here or in China, horses respond to the stimuli that we call aids in universal ways.

▶ This rider is prepared to use all of her aids as needed: seat, legs, hands, whip, spurs, and voice.

Using the Aids to Communicate

Your body movements convey dozens of messages to your horse, whether you want them to or not. To control your horse, you must be in control of your body and what it's saying. To understand how important that is, imagine trying to play music with a music critic standing next to you, criticizing every note: "You were a little flat on that last note. The tempo is too fast. No, now it's too slow. Your breathing is off …" Even if you're not a musician, you can see how terrible that would be.

As a rider, part of your job is to avoid a constant stream of criticism to your horse by keeping the messages from your aids consistent and correct. If you're not in control of your seat, legs, and hands, you're sending all sorts of distracting and confusing messages to him. The aids themselves should be short and intense to make them clear to the horse, but nearly or completely imperceptible to the eye of the observer.

INSIDE, OUTSIDE, LEFT, AND RIGHT

One of the things a new rider needs to understand is the terminology of inside vs. outside, which depends on knowing the difference between tracking left and tracking right.

The first, and simplest, definition is when you're riding around the outside edge of the arena. *Inside* is the side closest to the center of the arena. *Tracking left* means that you are riding with your left side (and your horse's) on the inside, toward the center of the arena. Your right side is *outside* or toward the rail or wall. When you reverse direction, you are now tracking right.

Any movement that involves a turn also has an inside and an outside. While you're in a turning maneuver, *inside* refers to the inside of the turn. For example, in a circle, inside is closest to the center of the circle.

(See box on page 55 for more on tracking.)

The Unconscious Aids

When learning how to give correct aids, a rider is really learning how to correctly influence a horse. Some newer riders are amazed to find out that their horses can tell so much about what they're thinking and feeling, especially when the rider is feeling anxious or worried. But it shouldn't be a surprise— clenching seat muscles, pinching knees, and gripping hands provide pretty clear information to a horse that a rider feels tense, fearful, or lacks confidence.

A Spectrum of Aids

In the process of learning how to give clear aids, riders will use the spectrum of negative, neutral, and positive aids.

A NEGATIVE AID is presented when a rider blocks a horse's movement, perhaps by sitting too heavily or too one-sided, or with no sense of feel for what the horse is doing.

A NEUTRAL AID is presented when a rider is just following along, not blocking the flow of energy in any way, not inhibiting motion, and just being free and natural.

POSITIVE AIDS are ultimately *educated* aids. These are presented when a rider is able to consciously shift weight left or right, forward or back, to send information to a horse. This rider can use her legs to apply the right amount of pressure and at just the right moment. The educated rider uses hands to restrain and allow forward motion with thought, good coordination, a sense of feel, and empathy.

The weight aids come first, then legs, then hands, followed, if necessary, by any combination of whip, spurs, or voice.

#6 Seat and Weight Aids

UNDERSTANDING SEAT and weight aids is a lesson in the science and bio-mechanics of both you and your horse. Riders often find it hard to comprehend just how sensitive a horse's body is. How can a little shift from one seat bone to the other, moving perhaps a few pounds of your weight from one side to the other, possibly influence how a 1,200-pound (544 kg) horse moves?

To appreciate the sensitivity of a horse's body, think about what happens if a fly lands on his back. He not only notices the nearly weightless fly, but is able to twitch just the right patch of skin to make the fly go away.

Regardless of how petite you might be and how subtle the shift of your weight, it's still many times more than any fly. Your horse certainly feels you, and that little shift of your weight, even through the saddle, goes a long way toward influencing the way your horse moves.

Of course, you can't use your seat as an aid effectively until you've established it as a foundation, as discussed in chapter 2. Once your seat is well established, you can begin to work on understanding how to control and influence the horse through subtle shifts in your weight.

▶ Picture your seat aid as corresponding to the numerals of a clock, with noon as the pommel and 6 o'clock the cantle; 9 o'clock is your left seat bone and 3 o'clock is your right seat bone.

Using Your Seat Bones

If you're not quite sure where your seat bones are or how to find them, start by finding a hard chair. Sit down with your hands underneath you, palms down. You will feel your seat bones going into the back of your hands. Wiggle around a bit and you will quickly see how the shifting of weight onto and between the seat bones works.

Using your seat and weight as aids can be broken down into three basic categories:
▶ Increasing the weight on both seat bones
▶ Easing the weight on the seat bones
▶ Shifting the weight to just one seat bone

Many instructors use the idea of a clock face to help you understand how to use these three categories of weight aids. Imagine that you're sitting on a clock. Twelve o'clock is at the pommel (front) of the saddle, 6 o'clock is at the cantle (back), 3 o'clock is the right seat bone, and 9 o'clock is the left seat bone. Sitting in the middle of the saddle, with your weight balanced evenly, is to have a neutral seat.

Making Your Seat Aids More Effective

▲ INCREASING SEAT WEIGHT

▲ EASING SEAT WEIGHT

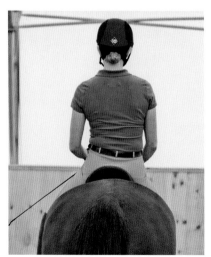

▲ SHIFTING SEAT WEIGHT

A well-schooled horse and a rider who knows how to use correct weight aids (combined with leg and hand aids) is a picture of what the FEI refers to in its section on the object of dressage, with the horse "achieving perfect understanding with the rider."

Effective aids depend on how the rider deals with the horse's constantly changing balance. As the horse's balance changes, he will influence you, and as your balance changes, you will influence him. Unless you're both frozen like statues, the balance relationship is always in a state of flux.

Here are some other important points to remember:

▶ To give weight aids, the rider needs body control.
▶ Weight aids, when correct, support the movement of the horse.
▶ In the classical context, weight aids are mainly used as forward-driving aids.
▶ Driving aids are always more important than restraining (hand) aids.

Increasing Weight on Your Seat

Shifting the weight of your seat to 6 o'clock by increasing the weight on *both* seat bones gives the horse a little *whoa* message. This weight shift, combined with a forward-driving leg (we'll talk about leg aids in the next section), is intended to encourage more activity from the horse's hind

legs. Shifting your weight back is used in half halts (see page 124), in all transitions, and when making a full halt. When you use your seat this way, the upper body stays tall, but you engage your core and "drop through" the seat.

Easing Weight off Your Seat

Easing the weight off the seat bones means shifting your seat to the front of the saddle, at 12 o'clock. Sometimes described as being "light and forward," this position provides a little natural *go* message from you to your horse.

An example would be easing the seat when coming out of a half halt, to match the *go* of the forward-driving leg. This easing of weight can also be used when warming a horse up, or when riding young horses, with the idea of being a little lighter on the horse's back.

A rider moving to 12 o'clock remains with the seat in the saddle, but there will be a little more weight on the thighs and in the stirrups. The extent of the lightening of the seat depends on how much weight is taken off the seat bones and put onto the thighs and stirrups. (See the discussion of light seat on page 25 for more.)

Shifting Weight to One Seat Bone

With this aid, the rider is shifting her weight either to the right seat bone (3 o'clock), which eases the weight on the left, or to the left seat bone (9 o'clock), which eases the weight on the right. Shifting to either seat bone results in a slight lowering of the hip and of the knee and moving more weight into the stirrup on the same side.

The movement is subtle, but this aid is critical for riding turns and bends using seat and leg aids, as opposed to using the reins as a primary steering aid.

Riders also use the seat aids in lateral movements and to indicate to the horse direction, tempo, and changes of gait. For example, a rider who is in working trot and plans to transition into canter while tracking to the left can use the 6 o'clock position for a half halt (preparing to canter), then a shift of weight, combined with leg aids, to roughly 10 o'clock, and the horse will know to canter on the left lead.

I find that learning how to use weight aids for canter departures is a thrill for students and a great display of what properly used seat aids can do. The first time a rider learns to canter off seat aids is usually followed by some response along the lines of "Oh, that's so cool!"

Your horse certainly feels you, and that little shift of your weight, even through the saddle, goes a long way toward influencing the way your horse moves.

Mistakes in Seat Aids

The Overly Heavy Seat

A common error in the seat/ weight aids is leaning back and sitting tensely, with the knees coming up and the lower leg shifting forward. Some riders take up this position in what they misunderstand to be a driving seat, which is actually grinding their seat bones into the saddle. The back should not go behind the vertical.

You will see this same incorrect image when a rider stays stuck in the 6 o'clock position, riding on heavy seat bones. (Remember, "heavy" in this context isn't to be taken literally. It refers to the movement of a rider's weight onto or away from the seat bones, not to her overall weight on the horse.)

Locked Hip and Pelvis

Another common mistake in seat aids happens with the rider who becomes stuck at the 12 o'clock position because of a locked hip and pelvis. Being stuck hollows the back and forces the rider to pinch at the thigh. Unfortunately, rather than giving the horse a natural *go*, this translates as *don't go* and inhibits the horse's motion.

▲ OVERLY HEAVY SEAT

▲ LOCKED HIP AND PELVIS

Leg Aids

"PUT YOUR LEG on that horse." "Keep the horse in front of your leg."

"Straighten the horse with your leg."

"Ride the horse between your legs."

In general, I find that riders know that the leg is part of how they communicate with their horse, but they don't understand the different ways to use the legs and the different messages they send. It's a bit like being in Spain and knowing how to ask for directions, but not knowing the words for right and left. So when an instructor uses various terms for leg aids, the rider is often confused as to what, exactly, she needs to "say" to the horse for him to "hear" the aid.

In terms of importance, the leg plays second fiddle to the seat/weight as an aid. But just as in a musical group, the second fiddle still needs to be in harmony; your seat-and-weight aids always have to work as a coordinated team with the leg aids. The rider with the most effective leg aids is the rider whose thigh, knee, and lower leg are supple and draped passively around the horse except when asking for something. The correct use of the leg does not disturb the rider's position; the action is quick and then the leg returns to passive.

DON'T BE A NAG

Your horse must respect your leg aids. The key to being effective is not to nag the horse with legs that are churning like eggbeaters. The rider who is constantly nagging the horse will make the horse "dead to the leg." Just as a mother learns to tune out a teenage daughter who is constantly complaining about something (and vice versa!), the horse will stop listening to a leg that is never quiet.

I once heard Lendon Gray (two-time Olympian, founder of Dressage4Kids, and CEO of The Dressage Foundation) describe this problem with insight. As she so aptly put it, "The leg means *go* and the hand means *whoa*." This is a great way to think of how the horse *should* respond to a rider's leg.

Unfortunately, what often happens instead is that a rider uses her leg to ask a horse to go once, twice, three times, and finally, on the fourth try, the horse actually moves forward. When this is repeated time and again, the horse learns to move off the fourth ask. If we ask politely once and get no answer, the horse has to hear *clearly and promptly* about the mistake from the rider (which might include the use of whip and spur). A few times of insisting on a prompt response and a horse will be trained to move off the first ask *and* keep going. In the development of good leg aids, you also need to develop a sense of feel for when to ask and when to stay quiet. Being passive with the leg – at the right time – is just as important as being active.

If you are using your leg correctly, but are still having problems getting the desired response, the problem may be in the timing of your leg aid in relation to your other aids. A good forward-driving leg needs to be in sync with your weight and hand aids, and with the horse's motion. For example, if you are giving a correct forward-driving leg for a trot transition but your hand is too strong and in a restraining mode, and perhaps your seat is in a holding position, your horse will have trouble answering your leg with 100 percent enthusiasm.

Using the Correct Leg Aids

You can break down the use of the leg into three basic modes:

▶ Forward-driving leg aids
▶ Forward-sideways leg aids
▶ Supporting/regulating/guarding leg aids

The rider must learn to adjust the various leg aids when using them at different gaits. Generally speaking, a walk needs alternating legs to confirm and continue a good walk, based on the fact that the walk is a four-beat gait. The trot, being a two-beat gait, is asked for and maintained with two-leg support, meaning two legs used at the same time.

The canter is different yet. The three-beat nature of the canter requires a rider to use one leg at the girth line and the other leg slightly back from the girth. This might seem a bit confusing for you as a rider, but imagine how confusing it would be to your horse if your legs were saying *canter*, but the rest of your aids were saying *trot*.

▲ FORWARD-DRIVING LEG AID

Recognize also that all horses respond slightly differently to leg placement and intensity of the aids. Learning to have a volume control on your legs is very important, and you need to adjust that volume depending on the horse that you're riding and what he feels like on any given day. I always encourage riders to start with the volume low, and dial it up if necessary. It's better than starting with your aids on full blast and having a horse respond with a full-blast reaction.

Forward-Driving Leg Aid

A forward-driving leg is the leg aid we use when we ask our horse to move. You should be able to give the horse a quick "pulse" with the lower leg and the horse should move forward in response. To be effective, the lower leg needs to be roughly 4 inches (10 cm) back from the girth, as this is the most sensitive part of the barrel. The pulse is most effective when the rider turns the toe slightly out so that the thigh can't clamp down. The leg needs to be otherwise steady, and the aid used thoughtfully.

An example would be when asking for a horse to transition from walk to trot. The leg asks for the trot in a quick action to the horse's side. Think of it as a pulse: on-off. The horse responds by moving into the trot. Because the horse had the correct response, the rider's leg returns to a quiet, neutral mode.

This is opposed to the idea that the rider's leg pushes into the horse for the trot transition and then stays there and continues to push to keep the trot going. Ideally, once you set your horse in motion and have determined the tempo, he remains at the tempo until *you* ask for an adjustment.

Common Problems with the Forward-Driving Leg

Giving a proper forward-driving leg aid can be quite difficult. One common error is the banging leg. The leg is going up and down, like the flapping wings of a bird who's trying — unsuccessfully — to take flight. There's minimum contact with the horse. Banging the legs against the saddle rarely has any effect because the rider, while physically active, isn't exerting much actual pressure on the horse.

A similar problem is when the leg is positioned too far forward. The result is that the rider

▲ A leg that's too far back is a problem. In this position, the heels come up, the knee pinches on the saddle, and the horse does not receive the message to go forward.

is mostly just kicking at air. You need to make contact with the horse to have any reaction.

Either of these errors can be combined with pumping the seat. This is when the rider rocks back and forth in the saddle, as if trying to rock a car that's stuck in the mud or snow. The upper body should be calm and steady, not shifting back and forth to make the horse's forward motion happen. Yes, this is a seat error as well, but it's a consequence of not using the leg aids correctly.

Forward-Sideways Leg Aid

The second way to use a leg aid is when you want to ask the horse for a forward-sideways movement, such as leg yield (see page 154). This aid asks the horse to step in a forward direction and to move

▲ FORWARD-SIDEWAYS
LEG AID

sideways at the same time, with the hind leg stepping under the body. Your leg should be slightly behind the forward-driving leg position. The sideways push comes from the leg that's opposite to the direction of the movement. For example, if you want your horse to move forward and to the left, the signal comes from the right leg. Timing, as always, plays an important role in the forward-sideways leg aid.

This aid is most effective when it's used the moment that the horse's hind leg comes off the ground on the side away from the intended movement. When asking for the forward-sideways movement to the left, put your right leg on the horse just as his right hind leg is coming off the ground. The leg should be applied as a pulse to the horse's side as opposed to a steady push.

Regulating, Guarding, or Supporting Leg Aids

The third way to use the leg as an aid is to support the horse's body and help guide him in a balanced way. The different terms used for this aid — regulating, guarding, and supporting — mean essentially the same thing. This leg aid is not as much an active asking leg as a leg that says (at the right moment) *okay, far enough* to the horse's body.

For example, if a horse is on a circle, the outside leg forms a wall of sorts to let the horse know that that's as far as his body should move out. If a horse is on

▼ REGULATING/GUARDING/
SUPPORTING LEG AIDS

a straight line down the wall, the outside leg supports the horse's body so he can't bump into the wall in a moment of questionable balance.

With the forward-sideways leg aid, the outside leg regulates just how much sideways movement you want. The leg is placed a bit back from the girth and used by putting pressure to the horse's side when necessary. Start with a light impulse on the horse and increase the intensity of the impulses until you get the response you need.

#8 Hand/Rein Aids

Too many riders ride with too much rein. It's not that we want to use the reins too much, but unless we are self-disciplined and correctly taught how to use the seat and legs, our hands seem to just take over the game. What many riders don't realize is that developing good hand aids, and therefore establishing a good sense of contact with or connection to the horse, is the most important step in a riding education.

There's a saying that every time you want to use your hand, use your leg instead. The point is to always keep the aids operating in the correct priority. The order is: first seat and weight aids, then legs, followed by hands. A novice rider almost always has the order backward: hands, legs, and maybe some seat aids. It takes a knowledgeable, conscientious, and skilled rider to operate in the correct order.

There are five ways to use the reins to communicate: asking (taking), yielding (giving), nonyielding, regulating, and sideways-acting. The simplicity of these words belies the challenges. With each method, you have the option of using the two reins together as the same aid, using only one hand as the aid, or using one hand as one aid and the other hand as a different aid. Yikes!

This question of which hand, when, and how is repeated hundreds of times in any given ride. Then, mix in your seat and leg aids! This is why coordination is so important.

The Asking and Yielding Reins

The asking and yielding reins take and give in the connection process, and you need to consider these two always as a pair; you can't ask (take) without yielding (giving), and you can't yield without asking (taking). There's no right or wrong about which comes first. It depends on what's happening at the moment.

As you hold the reins in your hands and send your horse forward with the driving aids of seat and leg, the giving and taking of the rein must be finely tuned, subtle, and sensitive. You will use the taking and yielding rein in all half halts, full halts, and transitions between the gaits and

▲ YIELDING REIN

▲ ASKING REIN

within the gaits. These rein aids are also used to improve your horse's contact with the bit.

A Subtle Movement

To more clearly understand what the taking (asking) rein is doing, think of it as a moment of more firmly closing the fingers, even to the point of making a fist. For a stronger effect, you can turn the hands *slightly* inward with a bending of the wrist.

The yielding (giving) rein happens with a softening of the hand (but not opening of fingers), or moving the hand a little forward by softening the elbow joint. This yielding aid can be so minute that some trainers refer to it as "the fourth-finger release". They're talking about a very small change — softening just your fourth finger (the one the rein is next to) on just the inside hand. This gives just a tiny bit of relaxation to the rein on that side.

A good rule when learning how to give and take (ask and yield) is to never take without giving.

The Nonyielding Rein

The second hand aid, the nonyielding rein, is a short hold (measured in seconds or split seconds), but without any pulling. This aid is used when you are having trouble with a horse who is resisting the contact (more on contact on page 95). A rider needs to be using seat, legs, and then the nonyielding rein to remind a horse to go forward into the contact.

As soon as the horse softens and moves forward without resistance in the hand, the rider softens too, relaxing the hand aids. You don't want to hold him back, so the timing here is critical and not an easy thing to master. The duration of the hold is dictated by the horse's response.

The Regulating Rein

As much as one leg becomes a regulating or guarding leg, one rein can become a regulating or guarding rein, the third type of hand aid. You'll want to use an outside rein as a guarding rein when the horse is being asked to bend or flex, as described in chapter 6. This regulating or guarding rein aid works in conjunction with the inside asking/yielding rein. On a circle or bending line, this outside rein may also be called the supporting or balancing rein.

The outside regulating rein is fairly firm, meaning you can really feel your horse solidly in the outside rein, but it still allows forward motion. This is opposed to the inside, asking/taking rein, which can also be called the softening rein, suppling rein, allowing rein, giving rein, or rewarding rein. This rein needs to be a more giving and softer rein than the regulating rein.

▲ For a nonyielding rein, both reins are held in a short, firm grip.

Sideways-Opening Rein

The last of the rein aids is the sideways-opening rein, sometimes called a sideways-acting rein. You won't see this rein aid used with any upper level horse, except in the rarest of circumstances. It is generally reserved for helping a horse figure out how to get somewhere, most often a young horse or a horse who's learning something new.

If a horse is having difficulty on a circle, the opening rein (meaning you actually open the rein away from your horse's neck) gives him a chance to see where you intended to go. Also, in the case of introducing lateral work, the opening rein helps guide the horse in the right direction.

A good rule when learning how to give and take (ask and yield) is never to take without giving.

▲ Regulating rein: Here, the right rein supports and balances the outside of the horse.

▲ The sideways-opening rein is useful in training situations.

Common Rein Errors

▲ The most common mistake with reins is keeping them too fixed, which hinders the horse from moving forward properly.

Assuming you are riding with correct contact, it is a truism that overuse of the reins causes a multitude of problems. (For more on contact, see page 95.) The most common mistake is not giving enough or just plain taking too much. In the case of the rider who has fixed hands (see page 20 for more on hand position), the horse is constantly blocked in his efforts to move forward.

Soft hands and an elastic elbow are needed to follow the motion forward; if those are not in place, the horse will just keep bumping up against the equivalent of a brick wall.

Don't Pull on the Reins!

Some riders get stuck in the non-yielding mode, and the horse answers by fighting against the rider's hand. The rider answers back by pulling on the reins. An ugly game of tug-of-war often results, and believe me, the rider *never* wins.

Instead of pulling, the correct answer to the problem is to yield the rein. This softens the contact and begins the process of taking and yielding again to regain and correct the contact between your hands and the horse's mouth.

◀ Pulling is not an option. A tug-of-war between horse and rider accomplishes nothing. To correct the problem, yield the rein and try again.

#9 Auxiliary Aids

MOST HORSES are pretty good at listening to the three main aids, but there are times when you might find that your horse is just tuning you out or not responding with the energy that you were hoping for. This is where auxiliary aids, also called secondary aids, come in. These aids are whip (or crop/bat), spurs, and voice.

Auxiliary aids, as the name implies, are a backup to your regular aids. For example, you should always ask your horse to go forward with your seat and leg first. If you find, however, that when you ask with your leg, your horse either ignores the request or is woefully insufficient in his response, it's not going to help much to ask with more leg. The best solution is to give him a quick tap with a whip or include a touch of spur with your leg aid.

Auxiliary aids make your horse more responsive to your regular aids and are there to reinforce the initial request. It's important to remember that carrying a whip and wearing spurs is not a license to punish your horse. The whip and spurs should be used briefly and sparingly.

The auxiliary aids are used to reinforce, never to punish.

▲ The three auxiliary aids are the whip or crop, spurs, and voice.

SELECTING A DRESSAGE WHIP

Dressage whips come in varying lengths and styles. You'll find them from 36 to 52 inches (91–132 cm), with lashes (the little flicky part at the end) that add from 1 inch (2.5 cm) to 5 or 6 inches (12–15 cm). (For showing, be sure to check with the current rules for length, as the rules seem to fluctuate and you wouldn't want to be disqualified for having a whip that's too long.)

The handles can be thick or thin. If you've got small hands, you'd be well advised to go for something on the thinner side. A knob at the end to keep the whip from slipping through your hand is always a good idea.

Using a Whip or Crop

In dressage riding, the dressage whip is the most common secondary aid and is used to get your horse's back legs going. There are also occasions when you might use the dressage whip on the shoulder to help reinforce the regulating aids or the forward-sideways aids. When not using the whip, you can rest it across your thigh.

It requires practice and timing to develop skills with the whip. A huge whack could cause your horse to blow up. And constant flips of the whip become nagging. The best way to get your message through to the horse is a brief, sharp tap. It's important that your horse not fear the whip, but at the same time have a healthy respect for it. It should never, ever, be intended to hurt your horse.

In what hand should you carry the whip? The best answer: carry it in the hand where you

▲ CARRYING A CROP

most need it, which depends on the strong and weak side of both your horse and you. The whip should be carried where you need to give your horse a message that isn't getting through with your regular aids. If your horse needs more help on the left, or if your left leg is the weaker of your leg aids, carry the whip on your left.

Practically speaking, however, when you ride in an indoor arena, a whip carried in the outside hand tends to scrape the walls and is not overly useful. Most dressage riders tend to carry the whip in the inside hand, at least when riding indoors.

For jumping, you'd carry a crop or bat. These come in all shapes and sizes, but generally a crop ranges from about 16 to 25 inches (40-60 cm) long and has a small popper or flapper at the end — a little leather or synthetic leather piece that makes a bit of noise when you tap your horse. A bat is in the same length range, but usually has a wider leather piece that makes more noise.

A crop is used on a horse's shoulder, and it is also acceptable to put the reins in one hand and use the crop behind the saddle pad. Aside from asking for more forward motion, the crop is used to support the outside rein and keep the horse's shoulder from falling out.

▼ CARRYING A DRESSAGE WHIP

Using Spurs

Some newer students view spurs as cruel and say they'd never use them. This feeling may hark back to the image of the cowboy, with his huge, pointy spurs hanging off the back of his boots. English-style spurs are nothing like that. Generally, they have a blunt end ranging from a small knob to a 2-inch (5 cm) neck.

Even blunt spurs can cause injury if poorly used, and the more aggressive styles should be used only by the advanced or professional rider who has the skill and knowledge about how to use them best — which is usually very sparingly.

After a student develops an independent seat and control of her legs, adding spurs as a secondary aid is helpful. The spur's purpose is to help generate impulsion or enforce respect for the forward-driving aids. The spur reinforces the leg aids, and can be used separately from the whip or as a complement to it.

As with any aid, use of the spur should be thoughtful, brief, and promptly applied. When you're wearing spurs, be sure you've got a steady, controlled leg or you might get more than you asked for. As with the whip, using spurs too much is nagging the horse — sooner or later he'll just stop listening.

A word of caution: if you haven't used spurs and/or if you are working with a horse who has never been ridden with spurs, think twice before putting them on. You can always drop a whip if the horse is reacting negatively to it, but you can't drop your spurs if they're having a bad effect.

Using the Voice

The voice is the last of our auxiliary aids. As with the other auxiliary aids, the voice is important, but needs to be used with a light touch. Some people deliver a running monologue to their horse from the minute they get on until the minute they get off. I saw a great quote that said, "Talking riders don't ride." The voice should be used sparingly and thoughtfully.

▲ USING THE VOICE

The voice is very helpful in working with a young horse because it can offer reassurance. The tone is more important than the actual words. If a horse is scared or worried about something, a rider's calming, low-pitched voice can be comforting. Conversely, if a horse is being sluggish or inattentive, a quick click can be helpful. Sometimes, you might even hear a rider "growl" deep in her throat — *grrr* — to make the horse listen and be more obedient.

The voice aid must be exercised with caution. Although it can be handy in working with young horses, more experienced horses have often become quite dull to it. In addition, if you aspire to show your horse, you should be aware that use of the voice is not permitted in a dressage competition, making the voice a training aid only.

▼ USING SPURS

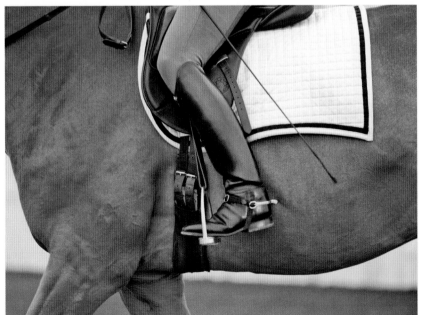

#10 Diagonal Aids

NOW THAT THE BASIC AIDS have been established, we can put the icing on the cake: diagonal aids. These are the marriage between your inside and outside aids. In the beginning of a rider's education, aids are used in the most basic form, meaning just on one side of the horse. An example would be if you wanted to turn right by using your right rein. In this early stage there is no crossover of aids between the two sides of the rider's body or the horse's.

When riders are first getting a handle on how to steer and ask their horse to go forward, and when young horses are first in training, these lateral aids are simpler for both rider and horse. As the abilities and coordination of the rider grow, and

▼ The diagonal aids ask the horse to be balanced from the rider's inside leg into the outside hand (here the right leg and left hand).

assuming the horse is trained to the necessary level, she is able to develop the skills to use diagonal aids. Diagonal aids are used simultaneously on opposite sides of the horse, such as the left leg and right hand when tracking to the left.

Using Diagonal Aids

When you are able to ride using an inside leg to balance the horse into an outside supporting rein, you are using diagonal aids. Being able to ride with these aids in place is something like graduating from middle school and moving on to high school. The basics are established, and now you can use your solid foundation to get to more advanced riding. Diagonal aids are part of that next-stage work.

Let's break down what happens when you use diagonal aids. Picture yourself riding on a circle. You have your horse balanced into both reins and moving with good forward energy.

YOU START WITH YOUR WEIGHT AIDS, in this case by placing a bit more weight on your inside seat bone.

NEXT, YOUR INSIDE LEG ACTS AS A FORWARD DRIVING AID, pushing lightly at the girth area to encourage your horse to step more forward with the inside hind leg.

YOUR OUTSIDE LEG OPERATES AS A GUARDING LEG, meaning it's positioned slightly back and used with light pressure on his side. You use the guarding leg to

▲ Here the horse is flexed to the left with the inside rein and balanced using diagonal aids between the left leg and right rein.

establish a "wall" that limits your horse so that he won't make the circle bigger or let his haunches swing out.

NEXT, THE INSIDE REIN SOFTLY AND SUBTLY ASKS for a slight flexion to the inside so that you can just see the horse's inside eyelashes. (For a better understanding of flexion, see page 118.) The outside, regulating rein allows the horse to be flexed and bent to

the inside without permitting too much bend in the neck and head.

Although I've described the aids in order of use, when you're actually on the horse, the timing is measured in split seconds, and the aids are applied as you feel the need to either adjust or maintain how your horse is moving. There is no cruise control with a horse, so your aids can't be thought of as something that you set up and then leave alone.

BEND AND FLEXION

The topic of diagonal aids is intimately tied with lateral bend and flexion. For a horse to be ridden on the diagonal aids, the rider must establish proper flexion and bend. For more on this subject, see chapter 6. For now, we'll just say that flexion is when we are able to flex our horse slightly at the poll so that we can see the edge of our horse's inside eye. Bend is our horse's ability to bend through the entire length of the spine, from head to tail. You can have flexion without bend, but you can't have bend without flexion.

CHAPTER 4

WORKING ON GAITS

Anyone new to the horse scene would be mystified trying to sort out the differences between a horse's movements if he should pace, tolt, rack, jog, trot, lope, piaffe, or gallop! Thankfully, in English riding there are just three basic gaits with which to work — walk, trot, and canter — each with a clear definition and purpose.

THE THREE BASIC GAITS come from a horse's natural movements. A horse in his natural environment will use any of these three gaits freely and at a moment's notice, although he will almost always walk unless there's a reason to move faster. In a field when there's no pressure on him, he walks from one grazing spot to another, and walks to find water or shade.

He will pick up a trot for a variety of reasons: to show alarm, to escape from pesky flies, to play with friends, to avoid being caught, and to show interest in a human who's got food.

The canter typically expresses excitement of one kind or another. For example, horses may break into a canter when startled or when a new horse is introduced to the paddock.

EQUINE ATHLETES

Although any reasonably fit horse with decent conformation can perform well in English disciplines, many English-style riders gravitate toward a few particular breeds that excel at dressage, jumping, and cross-country because of their natural abilities. Year-end championships across the country (and around the world for that matter) are rife with European warmbloods such as Hanoverians, Oldenburgs, Holsteiners, and Dutch Warmbloods. You'll also find a large number of Thoroughbreds (especially in eventing) and Thoroughbred crosses in all the disciplines.

Traditionally, these breeds or crossbreeds are horses who have naturally good gaits and an athletic inclination, which is why they are sometimes referred to as "sport horses." The term describes a *type* of horse — one with good movement — and not a particular breed (although there are exceptions where the phrase is part of the breed title, such as the New Zealand Sport Horse and the British Sport Horse). Sport horses have long, athletic strides and the ability to move through their whole body. They also tend to have good minds (meaning steady and calm under pressure and stress) and a good work ethic.

There is the occasional disparaging opinion about the warmbloods: sometimes people call them "dumb bloods" because of the minority of horses of this breed type who are so slow on the uptake that they're thought of as just big, dumb horses. Remarks like this probably come from riders who enjoy the quick mind (and maybe unsteady nerves) of a hotter Thoroughbred.

Rhythm and Regularity of Gaits

The words "rhythm and regularity" should be ringing in your ears every time you watch a horse move. You can tell a lot about a horse's gait by watching him move without human influence.

Is he inclined to shuffle along in the walk, dragging his toes? Or does he march along like he's just signed up to be a Marine? Is the trot a little jog or does he lift his legs high and show expression? Does he canter in a gentle lope or with energetic leaps and bounds through the air?

When looking at a horse move in any of the three basic gaits, the purity of the movements is the key, and rhythm and regularity are at the heart of it.

For English sports (dressage, jumping, eventing), the ideal horse has good rhythm in his gaits and a natural way of moving. As with people, some horses are simply more athletic and better "movers" than others. Few people are an incredible athlete of Michael Jordan's caliber. Similarly, not every horse has what it takes to make it to the Olympics.

Whether your horse is a gifted natural mover or needs help with all of his gaits, you need a fundamental understanding of each of the three gaits and all their variations.

A horse in his natural environment will use any of the basic gaits freely and at a moment's notice, although he will almost always walk unless there's a reason to move faster.

AT LEFT: The three basic gaits (walk, trot, canter) come from a horse's natural movements.

#11 The Walk

IF THERE'S ONE POINT THAT'S misunderstood by the vast majority of riders, it's the importance of the walk. Unfortunately, riders all too often come into the arena and just walk around for a while before getting to the other stuff: trotting, cantering, jumping. Walking is viewed as a means to an end without any value in and of itself. Wrong!

The official FEI definition is deceptively brief: "The walk is a marching pace in a regular and well-marked four-time beat with equal intervals between each beat. This regularity combined with full relaxation must be maintained throughout all walk movements." Notice that rhythm and regularity are part of the definition.

Note that when the FEI uses the term "pace" it is referring to the actual "gait," as in walk, trot, or canter. In U.S. terminology, however "pace" refers to the specific variations within a particular gait (i.e., working trot, medium trot). It also can mean an error in the walk rhythm.

▶ The walk may seem like the easiest gait to master, but it is the cornerstone of all gaits.

The Importance of the Walk

One look at any dressage test and you'll see how the governing bodies of horse sports value the walk. In virtually every dressage test, walks are weighted heavily in the scores. Starting with the basic U.S. Dressage Federation (USDF) Intro tests (just walk and trot), in every USEF test, and up through the FEI Grand Prix level, walk is given a coefficient of 2.

If you're not familiar with dressage test scoring, a coefficient means that a given score on a movement is multiplied by that number. So a score of 5 on a free walk is multiplied by 2 to give 10 points. The scoring indicates that a walk is twice as important as an "ordinary" movement on the test. The quality of the walk can win or lose dressage competitions.

But it's not just competition that should concern you. The walk gives you the chance to work on all the same skills that are used in the higher gaits, including aids and balance. I like to tell students that if you don't have whatever you're working on in the walk, you're not going to have it in the trot, and if you don't have it in the trot, you're not going to have it in the canter. So establish everything in the walk first.

Variety in the Walk

Four varieties of walk are recognized in classical riding: medium, collected, extended, and free. For the lower levels of training, you should focus on the medium and the free walk, and those are the two we'll explore here. (Collected and extended are explained for your knowledge, but they are reserved for the more advanced horse and rider.)

There's one more consideration. The FEI recognizes and describes the free walk, and it is used in competition in the United States, as well as in the United Kingdom, New Zealand, and elsewhere. But the Germans, who are the source for much of what we define today as classical riding, don't recognize a free walk in competition. They use the free walk only in training.

Medium Walk

The medium walk (considered the horse's natural walk) is the primary walk that all horses should have. This means that your horse should be moving under saddle the same way he does when walking freely in a field. To be in a correct medium walk, the horse needs to maintain a light, soft, and steady contact with your hand.

The key characteristic of a good medium walk is overtracking. The hind legs land ahead of the marks from the front feet so that the horse is actually making hind hoofprints in *front* of the ones made by his front legs. A technically correct medium walk overtracks by a distance of the length of two hoofprints.

Students often ask the reasonable question, "Well, how am I supposed to know whether he's overtracking if I'm sitting up here? I can't see his feet." Here are some clues: in a medium walk, you feel your horse gently moving your hips and seat in a steady, rhythmic manner. You'll notice his head and neck going up and down and forward and back as he reaches through his back and balances himself through the natural movement.

In training, mirrors can help you identify the feeling of a correct medium walk, but there's no substitute for eyes on the ground. Take advantage of other riders in the arena and your instructor to check whether what you're feeling is, in fact, a medium walk.

Free Walk

A free walk is a focused walk without the steady contact of the medium walk. You allow the horse to move along on a longer, looser rein so he can relax and stretch down and out with his neck and head, without the firm connection with your hand. The FEI says this exercise "gives a clear impression of 'throughness' of the horse and proves balance, suppleness, obedience, and relaxation." Unless a rider takes the time to explore a free walk with the horse in training, she may find that, in competition, her mount tends to get lost, look around, and wonder whether this is the end of the ride.

One of the hardest parts of the free walk is returning to shorter, more connected reins. You want your horse to accept the contact without bracing at the poll or resistance in the mouth. Ask any judge who sits at letter C viewing the lower-level tests, and he'll tell you they can almost count on seeing resistance at this point in the test. The transition from free to medium walk needs to be part of our daily rides.

▼ MEDIUM WALK

▼ FREE WALK

ADVANCED WALKS: COLLECTED AND EXTENDED

Unless you're showing up through the levels, the extended walk and the collected walk should not be of great concern to you. At this point, you should just have an understanding of these walks so that you don't inadvertently ask your lower-level horse to do something beyond his capability or misinterpret what your horse is doing.

The main point of the collected walk is the marching nature of the movement. A hallmark of a correct collected walk is that the horse remains on the bit, with his neck raised and arched. The hind legs are engaged and all the legs are lifted higher because the joints are bending more than in a medium walk. A good description for the overall look of a collected walk would be "shorter and higher."

In an extended walk, the horse covers as much ground as possible without hurrying and without losing the regularity of the four-beat gait. Each stride is longer than that of the medium walk, but the rhythm is unchanged. To be technically correct, the horse in an extended walk should overtrack by three hoofprints. (Remember, a horse in a medium walk overtracks by two hoofprints.) The horse stretches out his neck and head while maintaining contact. The horse's nose must be clearly in front of the vertical.

Prancing Isn't Collection

It's worth mentioning how often someone without an educated eye will mistake a tight, tense horse for one in a beautiful collected walk. A hot, nervous horse might prance prettily in the walk, but this is the polar opposite of a true collected walk because the dancing, active motion comes from tension or excitement, not from the composed energy that informs a proper collected walk.

Riders are often confused about collected walks because the term "collection" is used differently for walk than for any of the other gaits. For all other gaits, collection needs impulsion, and impulsion, by definition, requires suspension, which is when all four feet are off the ground at the same time.

But the walk doesn't have any moment of suspension, given the four-beat nature of the gait. Logic would tell you that if you don't have suspension, you can't have impulsion, so a true collected walk is impossible. The walk, however, is the gait that breaks the rules. Although it's not technically impulsion that you have in walk, since there's no suspension, your horse should give the *impression* of having forward energy delivered by the hindquarters.

▼ COLLECTED WALK

▼ EXTENDED WALK

Problems in the Walk

So just what's so hard about a walk? It turns out that a good walk is one tough gait to get and to hang onto. It's often said that the walk is the easiest gait to ruin and the hardest to preserve. Some horses are born with a naturally good walk and, if this is the case, try not to interfere with what Mother Nature so graciously handed out. Horses who are born with a walk that is challenged will, in all likelihood, remain challenged despite good training.

The most serious mistake in the walk is the loss of rhythm and regularity. This loss of rhythm can be broken down into two major categories. The first is short-long walking, when one limb takes a shorter stride than the other. It can almost look as though the horse has a slight limp in his walk.

Another common problem is when a horse begins to pace, which is when two legs on the same side move almost simultaneously, and the correct four-beat walk turns into a two-beat walk.

Recognizing Walk Mistakes

There are both audio and visual clues that your walk is in trouble. Listening to the footfalls on a hard surface, you should clearly hear four beats. If what you hear is more in the nature of a two-beat rhythm, you've got yourself a pacing walk.

Visually, if you're watching a walk from the ground or riding in an arena with mirrors, you should be able to see a V shape between the legs as the hind foot swings forward, almost touching the front foot.

A breakdown in the horse's walk rhythm is mostly caused by tension and crookedness, and often it's the nongiving hand of the rider that's to blame. You need to allow a horse to use his neck; if you don't, tension usually comes next. When the rider stops following the motion of the horse's neck with her hand, the horse gets stuck and tense. From that tension, the horse stops being loose and supple and the result is a bad walk.

Unfortunately, some riders lose track of the importance of walk — the most basic gait of all — as they rise to the higher levels. There has even been an effort by some international competitors to have the walk removed from the Grand Prix tests.

Perhaps it's a sad commentary on what happens to a simple walk when advanced riders allow tension to enter the picture as they push their horses for brilliance in the show ring, but a well-known upper-level rider was once rumored to say, "Of course my horse can't walk. He's a Grand Prix horse."

▶ At a good walk, the horse's legs create a distinct V shape as the hind legs step up and nearly touch the front legs with each stride.

Improving the Walk

The walk is a very hard gait to get right, and there are only a few tools for fixing it. If your horse doesn't have a good walk, the first thing to do is to check your seat to make sure you are moving freely through your pelvis and not blocking your horse's motion.

Then, check what your hands are doing. Are they quiet and are your elbows elastically connected to the bit? These two fixes will at least give him the freedom to move.

Your leg aids can help the walk. The gait needs one-leg support, meaning one leg at a time

can softly push against his side to move him along, but you need careful timing of the aid. As your horse walks, his rib cage naturally swings from side to side. As it swings from left to right, push a little with your left leg to make the swing, and consequently the stride, a little bigger. The same can be done with the right leg. But limit the aid to a few strides; a constant push of leg after leg will nag the horse.

Sometimes, just leaving the walk alone for a while and working in other gaits gives a horse with a tense walk a chance to become more relaxed.

WHAT DOES "TRACK" MEAN?

Confused by the multitude of ways the word "track" is used in the horse vocabulary? A racetrack is pretty obvious, and we've already explained track left and track right, but here's a cheat sheet for some other tracking terms:

The track is the place where you ride on the outside perimeter of the arena on a space that's next to the wall. This can also be called the wall or the rail.

Sometimes you'll ride your horse on the *inside track*, which is one horse width closer to the center of the arena than the main track.

Tracking up means your horse's back feet are stepping far enough forward with his hind feet to meet to hoofprints of his front feet.

Overtracking is where the horse's back feet overstep the prints made by the front feet.

(See box on page 32 for more on tracking.)

▼ TRACKING UP

▼ OVERTRACKING

If your horse doesn't have a good walk, the first thing to do is to check your seat to make sure you are moving freely through your pelvis and not blocking your horse's motion.

#12 The Trot

THE OFFICIAL FEI definition of the trot is a bit more complex than the walk definition, but the key point, once again, is rhythm and regularity:

"The trot is a two-beat pace of alternate diagonal legs separated by a moment of suspension. The trot should show free, active and regular steps. The quality of the trot is judged by general impression (i.e., the regularity and elasticity of the steps, the cadence and impulsion in both collection and extension). This quality originates from a supple back and well-engaged hindquarters, and by the ability to maintain the same rhythm and natural balance with all variations of trot."

The trot is the gait where we seem to spend most of our riding time. We trot in the arena, on the trail, when warming up, cooling down, practicing ring figures, and so on. It's a comfort zone for most riders and a place to work on position, aids, and the suppling of our horse.

The trot, as the walk, can be broken down into several different kinds, with detailed criteria for each: working trot, lengthening of stride, medium trot, collected trot, and extended trot.

▼ WORKING TROT

▼ MEDIUM TROT

▼ COLLECTED TROT

▼ EXTENDED TROT

Working Trot

THE WORKING TROT, which falls between the collected and the medium trots, is the most frequently utilized. It is a forward gait with even balance and rhythm, marked by elastic steps and good hock action. In the working trot, the horse's hind feet should at least cover the front hoofprints. A horse that is naturally active in the hindquarters will have an easy time getting the "engine" going (meaning the powerful energy source of the hind end).

A common misperception is that a horse moves with the energy of the front end and the back end simply follows along. In nature, that's actually quite often how a horse is moving. As soon as we impose ourselves on a horse's back, however, we are changing the entire equation. Because the horse wasn't really designed to carry weight on its back, part of the rider's job is to turn the natural gaits into educated gaits.

This change, which comes through correct training, will give the horse what he needs, physically speaking, to carry us along. An educated gait teaches the horse to move with lifting, thrusting power (energy) from the hind end. The result is a back that comes up so that we sit on supple muscles as he pushes through his body toward the bridle.

Fixing Mistakes in the Working Trot

Of the many errors that are made in the working trot, the most common is a lack of energy. Without energy, a horse just shuffles along, looking as though he's hardly moving. In fact, you might even call this trot a jog, which is fine for Western riding, but not what we're looking for here.

A good trot must be *forward*, which is a shorthand way of saying the horse must have energy, bigger strides, and a prompt reaction to the rider's leg aids.

Then there's the horse who is moving along all right, but shows no activity of the hind legs, which are strung out and sort of dragging behind him. As a consequence, this horse is what's called "on the forehand," meaning he's falling forward and using only front-end energy to move. The fix for this problem is a generous use of half halts (see page 124).

The hurried trot is another common error. Here, the horse is racing along, but his strides are short and choppy. The horse is running rather than stretching out his strides with his hind legs moving well under him and rounding up through his back. Again, half halts are the needed fix.

▼ At a good working trot, you can see the horse's hind legs coming well underneath him. His back is rounded up and he is reaching into the bridle with energy.

▼ Here, in contrast, he is all strung out, with his hind legs trailing, his back hollow, and his nose stuck out in front of him.

Lengthening Strides

As a horse develops athletically, he needs to learn to push off more through his hindquarters (see "Fundamental #23: Impulsion," page 102). As his ability increases, the trot strides cover more ground and actually get longer. This lengthening, as it's called, is the building block to the medium trot and eventually to the extended trot.

When you start to train for lengthening of strides, by default you also start to train for collecting strides, because you must give half halts to contain the horse's energy before asking for the lengthening, and again after the lengthening to return to a working trot.

Fixing Mistakes in Lengthening

Mistakes at this point in the riding program abound.

Lengthening needs to be understood as allowing contained energy to expand forward. Commonly, riders wrongly approach a lengthening as a chance to drive their horses to move faster, but speed, in the sense of a faster tempo, has nothing to do with lengthening.

What a rider wants is to cover more ground with each stride. The consequence of the horse making longer strides is that you move from point A to point B (or in the case of a dressage arena from F to X to H) more quickly.

Think of two soldiers, one with short legs and the other with long legs, marching in a parade. Their feet hit the ground at the same moment, but the shorter soldier has to lengthen his stride to match the distances traveled. Once again, half halts are the key to a fix, since they push the

horse to use his haunches for the source of the energy.

Running onto the Forehand

Riders may also let their horses "run onto the forehand," meaning that he is using his front legs to pull himself along rather than propelling himself with his hindquarters. These trots often look as though the horse is running downhill, as his center of gravity tips forward.

At this stage, riders will mistakenly use the hand to bring the horse back to a working trot instead of applying half halts with seat, leg, and then hand. Using too much hand results in a horse who braces and resists before returning to a working trot.

▼ Lengthening is about covering more ground with each stride, not moving faster. Notice the moment of suspension shown here, with all four feet off the ground.

▼ A horse that lengthens incorrectly will run onto his forehand, meaning that he is not using his hindquarters to propel himself. Notice how this horse looks as though he is moving downhill.

Medium Trot

A medium trot is the natural progression after lengthening stride in the trot. In this variation of the gait, the horse's steps are very clearly lengthened without giving the impression that he is hurrying. The rider allows the horse to carry his head a little more in front of the vertical, and the horse lowers his neck and head slightly. The medium trot is a true sign of the development of impulsion.

To develop a medium trot you need to develop the *elasticity* of the horse's topline. The topline refers to the side-on view of the horse's arch from poll to tail. The horse lengthens his topline by stretching and arching his neck as well as by rounding his back.

This is a demonstration of longitudinal suppleness, showing how easy it is for your horse to expand and contract his body from back to front, and from front to back. If you're watching from the ground, you'll actually see the horse get longer and shorter along his spine; the change is enough to be visible to a careful observer.

Fixing Medium Trot Mistakes

Until you've properly developed your horse's ability to lengthen his stride, you are not going to have a correct medium trot. Horses at this stage of training often exhibit characteristic mistakes that mirror errors made in basic training.

High on the list of mistakes is the horse who moves with his nose behind the vertical. This is caused by too much hand on the part of the rider; it can be fixed with a more relaxed and giving hand and elbow.

All Strung Out

Horses may also move with the nose above the bit, with hollow backs and front legs doing all the work, or some combination. These mistakes indicate a horse who's been allowed to "run" forward, as opposed to working through the back and producing energy from the increased use of his hocks and haunches. This horse is "strung out," meaning he's lost the elasticity of the topline and has lost the connection to both the rider's hand and his own back end.

This problem can be fixed by schooling with short intervals of the medium trot mixed with a return to the working or collected trot, which encourages the horse to stretch and contract his topline as he varies the trot.

A great thought to remember when working on this area of your riding: medium trots are developed, not discovered.

▼ A good medium trot comes from developing the elasticity of the horse's topline.

▼ With his hollow back and dragging hind legs, this horse cannot produce the energy he needs for a correct medium trot.

Collected Trot

▲ At a correct collected trot, the horse should look as though he is moving slightly uphill.

In the classical definition, the criteria for a horse moving at a collected trot includes remaining "on the bit" while moving forward, with the neck raised and arched and with the hocks flexed, engaged, and moving well under the body. When the horse engages his hind end and flexes his hocks, his croup lowers and his forehand comes up. This makes the horse appear to be more upright. The horse's strides are shorter than in the other trots; the impulsion, elasticity, and cadence remain consistent.

Fixing Collected Trot Mistakes

It's easy to spot a horse who is not in a good collected trot. The croup is not lowered, and the forehand appears flat or maybe even lower than the haunches. The back is braced and hollow, and the hind legs are left out behind the horse. Once again, this is usually a problem of too much hand.

Without a tactful and appropriate use of the right mix of driving aids and restraining aids, the horse will be unable to shorten his strides without losing the swing through his hind legs, and he'll lose the ability to take more weight on his haunches.

The driving aids need to be more prominent in the mix; you know you have it right when the back end stays active and you can move directly out of the collected trot into any of the other trot variations.

▲ Using too much hand and not enough driving aids often results in a hollow back and shortened strides rather than a correct collected trot.

Extended Trot

When we finally reach the stage of extended trots, we need to understand that the only difference between a collected trot and an extended trot is distance of ground covered. In the extensions, the horse is being asked to cover as much ground as possible, but in a well-balanced way.

This is the ultimate expressive show of impulsion, as the horse surges forward from the hindquarters. By classical dressage standards, the front feet should touch the ground on the spot where they are pointing and the movement of the front legs and the hind legs should reach equally forward. (See page 106 for a discussion of the current debate over "show" trots versus classical training.)

Fixing Extended Trot Mistakes

The mistakes in the extended trot are essentially the same as in the medium trot, just exaggerated by the movement. The horse's back becomes even more hollow. His front legs may fling out dramatically, but the back end is doing almost nothing. As with the medium trot, training in intervals mixed with other types of trot is very helpful.

▶ Mistakes such as a hollow back and dragging hindquarters are exaggerated at the extended trot.

▲ The lines show the equal angle of the legs as they step forward.

#13 Posting and Sitting Trot

THE BASIC ASPECT of the trot that can't be overlooked is how the rider deals with the two-beat nature of the gait. It seems obvious to people who have been riding for a while, but for the novice rider, how best to ride the trot is always a question. There's no doubt that at the walk we stay in the saddle, and for most purposes, the same is true for the canter. But for the trot, we have options.

The Posting Trot

One of the first things taught in English riding is how to post the trot. (The terms "posting" and "rising" are used interchangeably.) A posting trot allows the rider to balance on a horse and go forward with the motion.

The two-beat trot means that if we come up on one beat and sit the next, we miss the "bounce" portion of the gait, which makes the rising trot easier on both the rider and the horse. A rider needs to keep her feet under her body and in the straight line of head-hips-heels, even while posting.

In the early stages of learning to post, a rider often pulls her body up out of the saddle rather than letting the horse push it up.

When riding correctly, the horse's energy gives us a little boost to rise out of the saddle. I can always tell when a rider is using only her own energy if, when she asks for a trot, she starts to post, even if her horse hasn't yet started to trot.

Include Your Elbows

In posting, there also is an opportunity to better understand how your elbows need to be part of the game. When a rider is sitting down, the angle of her elbow is smaller. When the rider is up in the saddle, the angle increases.

How *much* the angle should change is based on many factors (how big the trot is, how short the reins are, etc.), but it is important that there be some change as the rider moves. This is critical to having elastic elbows (see Hand Position, on page 20) and developing a sense of connection to your horse's mouth.

Finding Your Diagonal

In the posting trot, we also need to pay attention to something called a "diagonal." This word is

▼ POSTING (RISING) TROT

There is a definite difference between being on the correct diagonal versus the incorrect diagonal. You can see it and you can feel it.

used frequently in riding, but in this case, it's a reference to the fact that a rider needs to be in sync with the horse's trot motion. There is a definite difference between being on the correct diagonal versus the incorrect diagonal. You can see it and you can feel it.

The correct posting diagonal is to be up (rising from the saddle) when your horse's outside front leg (the one closest to the outside wall) is forward, and down (sitting in the saddle) when his leg is back.

If it's not correct, you need to sit one extra beat, which turns out to be an even number of beats with your seat in the saddle. Sitting an odd number of strides, like three, does not change the diagonal and you'll continue to be out of sync with the horse's motion.

The Sitting Trot

The posting trot is used in the learning process to help us gain proficiency in controlling our body and controlling our horse. After these skills are established, you should transition to the sitting position. Learning how to sit the trot is part of what often turns out to be the long road to gaining our independent dressage seat.

Even after learning how to sit properly, we don't end our need for posting (rising) trots. All horses should be warmed up using the rising trot. Sitting too early, before a horse's back is warmed up, can cause tension and get in the way of a horse moving through a soft, swinging back. On young horses, most trot work should be done in rising until the horse is able to develop more strength in his back.

The first thing to keep in mind when giving aids for the trot is that because of the nature of the movement (two-beat gait of alternate diagonal legs), the trot needs two-leg support (using both legs as an aid at the same time) from the rider. This is an important difference from walk, which works with one-leg support.

To prepare a horse for a forward transition into the trot, a rider uses all of the aids: seat, legs, and hands, and perhaps might need some help from the whip or spur.

• To start, the horse needs to know that a transition is in the making, so use half halts to prepare him for what's coming.

• Then soften the connection ever so slightly so that the horse gets the sense that the "front door" is opening (as opposed to the rider who pulls back on the reins or sets her hands when asking for a forward movement.)

• Nearly simultaneously, use both legs as a forward-driving aid and adjust the seat aid to be lighter to encourage positive forward motion.

▼ SITTING TROT

#14 The Canter

JUST AS WITH THE OTHER GAITS, rhythm is the key to a good canter. Here is the FEI definition: "The canter is a three-beat pace . . . that is always with light, cadenced and regular strides that have a moment of suspension when all four feet are in the air. The quality of canter is judged by the regularity and lightness of the steps and the uphill tendency . . . acceptance of the bridle with a supple poll and in the engagement of the hindquarters. The horse should always remain straight on straight lines and correctly bent on curved lines."

As with the walk and trot, there are several variations of canter — working, lengthening of strides, medium, collected, and extended — but the most fundamental part of the canter is finding the correct lead (see page 72).

▼ WORKING CANTER

▼ MEDIUM CANTER

▼ COLLECTED CANTER

▼ EXTENDED CANTER

Working Canter

The working canter is home base for all of your canter work — a place to school your horse and a departure point for the other variations of canter. On the spectrum of canters, it's located between the medium canter and the collected canter.

The criteria for a good working canter are that the horse remains on the bit and goes forward with even, light, and active strides. The signs of a quality working canter are a horse who has an uphill tendency and is jumping forward. An uphill canter means that the horse's forehand lifts up as he springs off his outside hind leg.

In biomechanical terms, the first phase (the three-beat canter has a total of six phases) actually has the horse balancing all of his weight (and yours) on his outside hind leg before he jumps forward. By bending all of his hind leg joints, especially his hocks, the horse should spring forward as far as possible, with his hind legs moving toward his center of gravity. In the sixth phase, the horse has a moment of suspension, when all four legs are off the ground at the same time.

The more the horse jumps through in this moment of suspension, the better the canter. To get this kind of quality canter and maintain it, always think of each completion of the three-beat/six-phase sequence as a chance to ask for a new strike-off (or beginning) into canter. This way you are always refreshing your aids and therefore refreshing your canter.

▼ A good working canter

Fixing Mistakes in Working Canter

A good working canter requires a rider to pay close attention to how her horse is moving. The question that should be in the forefront of your mind is, "What are the back legs doing?" All too often, a rider just allows her horse to canter around on his forehand. Because he's pounding his front legs forward while his back legs are well out behind him, his strides are all strung out.

Riders make another common mistake when they push a horse to move faster in an effort to keep him from falling out of the canter. Speed doesn't really address the key problem here. The horse is falling out of the gait because his engine (back end) is not moving under his body to spring him forward. Riding the canter with half halts will keep the horse connected and engaging the hind end.

Cross Cantering

Sometimes a horse either cross canters or goes into a four-beat canter. A cross canter, also called being disengaged or disunited, is when the horse is on one lead in the front legs and the opposite lead in the back. If you're on a horse who is cross cantering, you'll feel as though you're riding a washing machine on the agitation cycle. This situation is most often the result of a horse who is out of balance or lacks coordination.

When a horse is cross cantering, you shouldn't just let him keep going. He needs to be brought back to a trot and then asked to canter again. If this is a frequent issue, you'll probably need to take a more serious approach to determine what the core issue is with his balance and coordination, including perhaps a visit from a veterinarian.

Four-Beat Canter

In the four-beat canter, the diagonal feet are not stepping simultaneously and the moment of suspension is nearly nonexistent. This serious canter fault often results from too much pushing by the rider, combined with too strong a hand. It can also occur when the rider is unbalanced and out of position. For a fix, perhaps getting up in a light seat and letting your horse freely move forward in the canter will help.

Test your horse out in the open. If your horse canters without the four-beat gait while in the fields, chances are you're lacking relaxation (looseness) in your arena work and need to further address the suppleness in your horse before cantering.

▼ This canter is a mess: the horse is braced in his neck and strung out behind, while his front hooves pound along in front.

Canter Lengthening

On the way to developing medium canters, much like medium trots, a rider needs to take the interim step of working on lengthening of stride. For the horse who's heading toward a career in the jumper ring or as an event horse, lengthening and collecting the stride will be imperative, as they will need to shorten or lengthen their strides to hit the right distance to a jump.

Recall from a previous fundamental (see page 58) that lengthening is the athletic development of the horse's longitudinal suppleness. Canter lengthening is based on the same idea. You are asking your horse to move with more ground-covering canter strides and then, without tension, return to a working canter. The quality of this work depends directly on the quality of a rider's half halts.

▲ Here is a clear, forward, uphill lengthening of the canter stride, shown by the angle of the line from croup to poll.

Fixing Mistakes in Lengthening

Speed is not the name of the game, but it's often the unintended result when a rider asks for lengthening and consequently the horse runs onto his forehand. As with the other gaits, too much hand is the most common error.

The rider who wishes to return a horse to a working canter and becomes heavy with the hand will make her horse tight in the back as well as tense and bracing in the neck and head. To move to a working canter correctly, you need to give the aids in the proper order of seat, leg, and then hand.

▶ This canter lengthening is heavy on the forehand and strung out behind, greatly changing the angle of the line.

Medium Canter

Just as in the trot, the medium canter is not discovered; it is developed. The medium canter falls between lengthening the stride in canter and the extended canter. In a medium canter, the horse is allowed to move the head a little more in front of the vertical, and at the same time to lower his neck and head slightly, although the poll remains the highest point.

The word "engagement" (see page 114) becomes very important in this part of training a horse. A rider really needs to be working on riding the horse up from behind by using half halts. The body of the horse is developing the ability to become like a spring.

Fixing Mistakes in Medium Canter

Commonly, you'll see riders leaning back when moving into medium and extended. This error leaves the rider behind the horse's movement. If a horse has been well prepared for the increase in impulsion and pushes through the hind end, it's helpful if you sit lightly forward, as in shifting your seat slightly toward the 12 o'clock position (see page 33), and take the weight off his back slightly, even if it's just for the first few strides. This adjustment in weight aids makes it easier for the horse to step forward in a balanced way.

Another common mistake is when riders ask for a medium canter across the entire long diagonal or long side of a 20×60-meter arena. To be developed correctly, the horse should only be asked for medium canters over short distances with frequent transitions. This training approach elasticizes the horse's topline and helps him develop the propulsive power that needs to come from the haunches.

TOP: In a correct medium canter, the horse is springing off his haunches and moving uphill.

BOTTOM: This horse is tense and stiff, with his nose carried behind the vertical.

Collected Canter

If the medium canter is when your horse starts to develop the ability to move like a spring, the collected canter is the full coiling of that spring. In the collected canter, a horse's strides are shorter and the movement feels and looks as though the horse is "rocking" as he moves.

It's important that the horse remains on the bit and moves forward with his neck raised and arched. This opens the door for the shoulders to move with greater mobility and allows the horse to have two important attributes: self carriage and an uphill tendency.

Fixing Mistakes in Collected Canter

From reading about the gaits in the order of walk, trot, and now canter, you can probably guess what the primary source of problems will be for the collected canter. Too much hand makes a horse brace in the neck and poll and cuts off his ability to truly engage with the hind end. This leads to a tight back, a high croup, and, once again, no real athletic use of the haunches.

TOP: In the collected canter, the engagement of the haunches and the uphill, energized strides are even more evident than in the medium.

BOTTOM: Too much rider hand is causing this horse to brace in his neck and poll and lose the energy that should be coming from his hind end.

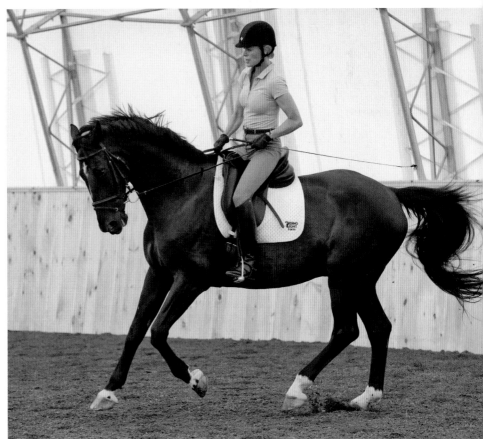

Extended Canter

The extended canter is the most ground-covering gait in which to work your horse. The horse is expected to spring forward with huge, vaulting strides, but he's not to hurry. He remains calm, light, and straight. As energized as the extended canter is, the whole movement should be well balanced, and the transition to a collected canter or to a trot should be smooth as silk.

To truly appreciate the beauty of this movement, watch a video of one of the world's elite riders. Seeing a horse jump forward in the canter with powerful engaging strides and then immediately transition to the collected canter is breathtaking. This smooth, supple transition is because the impulsion created in the extension is maintained through the transition to the collected canter.

Fixing Mistakes in Extended Canter

The energy and difficulty in performing an extended canter across the full diagonal of a 20×60-meter arena are tremendous. One of the first things that is often lost when working on extensions is a horse's looseness (see page 90). Excessive tension can result in loss of rhythm; a tight, hollow back; throwing of the hindquarters; and bracing with the neck and head. An extended canter should not remind you of being on a galloping Thoroughbred at the Kentucky Derby. The horse covers more ground because of the powerful thrust coming from the hindquarters, not because he's racing.

The correct way to train for the extended canter is frequent transitions between medium canter and extended canter. Throughout these transitions, you need to pay considerable attention to making sure you are increasing the driving aids in the downward transitions and not increasing the use of the hand. That might sound contrary, but you actually need more driving aids to make the horse swing through his back as he takes increased weight onto his hindquarters.

▼ Here you can see the clear moment of suspension that comes from the full, vaulting strides of a good extended canter.

▲ With his tight head, neck, and back, this horse is unable to drive through his hindquarters correctly. Notice the difference between these two photos in the rider's position and the expression of the horse's ears.

AIDS FOR CANTER

Much as there are three rules for a successful retail business (*location, location, location*), there are three rules for a successful canter: *prepare, prepare, prepare*. The quality of the trot really does determine the quality of the canter. A rider needs to take a high-quality trot and prepare the horse for canter by using one or more half halts. The half halts bring the horse's hind legs farther forward, toward his center of gravity. This gives him an easier way to strike off into the canter.

Here is a sample sequence for correctly moving your horse into a right-lead canter from a trot:

• Start with a good trot and prepare for the transition with one or more half halts.

• From a heavier seat (both seat bones weighted), move your weight onto the inside seat bone, a bit forward (weight the 2 o'clock seat bone if moving on right lead).

• At the same time, push your inside leg (right leg) against the horse at the girth line, move the outside guarding leg slightly behind the girth, and give a quick impulse with the outside leg.

• Your hands ask the horse to flex to the right, using the inside rein as an asking and yielding rein, and the left rein as a regulating rein, controlling the amount of flexion and keeping the horse's left shoulder from falling out.

• The moment the horse moves off into canter, the right (asking) rein must yield to let the horse canter through his body. From there, the seat and leg aids are used to keep the horse forward and the canter moving.

#15 Canter Leads

HAVING EXPLAINED all the variations in canter, one element is fundamental to all of them, and that's the canter lead. During the three-beat canter, one lateral pair of legs reaches further forward, indicating that horse is either on the left lead (a left canter) or the right lead (a right canter). A canter lead functions much as the posting diagonals of the trot, which help to keep you and your horse in balance. Using the correct lead in the canter means a world of difference for your horse's balance.

Defining Canter Leads

To be on the correct lead, when the horse is tracking right, the right (inside) front leg is always leading or slightly ahead of the outside front. When you change direction, you usually want your horse to pick up his left lead, meaning the left leg is always slightly ahead of the right. To be on the correct lead is to be in a true canter.

If you intentionally pick up the left lead while tracking right, it is called a "counter canter." If you pick it up in error, however, you're on the wrong lead.

▼ This horse is on the correct (in this case, the left) lead.

HOW TO GET A GOOD CANTER

In my experience, canter tends to be viewed as the big test of whether a person has really "arrived" as a rider. Before they reach that stage, most riders feel as though they're just learning the skills they need so that they can eventually canter. In some ways this is true, but by building up the canter into monumental significance, riders often bring tension to the table that doesn't have to be there.

The best approach to learning how to canter is to have it be part of your learning process while on the lunge line. Experiencing the canter while not having to worry about where you're going and at what speed gives your body a chance to deal with this new three-beat motion.

When you are off the lunge line, you need to make sure you are spending time clarifying your position and aids in walk and trot, and becoming an effective rider by riding your horse's gaits with consistent rhythm and regularity, good transitions, and half halts. The canter will be a natural progression of your riding skills.

▼ This horse is on the wrong lead and is clearly out of balance.

#16 The Halt

AN INSTRUCTOR ONCE TOLD A RIDER that her horse's best gait was a halt. Any horse owner would feel the "ouch" of that comment, but there could be a grain of truth to it. A good halt is important, and there is a whole lot more to bringing a horse to a standstill than slamming on the brakes. In addition to saying that the quality of the paces before and after the halt reflects on the quality of the halt itself, the FEI has a significant checklist of requirements:

"At the halt, the horse should stand attentive, engaged, motionless, straight, and square with the weight evenly distributed over all four legs. The neck should be raised with the poll as the highest point and the nose line slightly in front of the vertical. While remaining 'on the bit' and maintaining a light and soft contact, the horse may quietly chew the bit and should be ready to move off at the slightest indication."

Why Is the Halt Important?

The flippant answer about the importance of the halt is that you can get good scores on your dressage tests. While that is true, of course, a dressage test should be a mirror of your daily training, and a good, correct halt shows that your horse is working with balance, straightness, and correct contact. A good halt comes not by luck, but rather through thoughtful, consistent, correct training.

How to Make a Good Halt

The FEI describes the correct way to halt as "...the displacement of the horse's weight to the hindquarters by a properly increased action of the seat and legs, driving the horse toward a softly closed hand, causing an almost instantaneous but not abrupt halt."

Take note that, once again, the order of the rider's aids is seat, legs, and then hands. A heavy-handed rider who tries to stop a horse with force will not get a good halt. You need to start by preparing your horse for the downward transition by using half halts (see page 124). As the horse begins to gather himself up, you close your seat a bit more, close your leg a bit more, and then close your hand a bit more. Use the aids in this order, but nearly simultaneously.

If you lighten your hand aids (think about just "stopping riding") at the moment of completion of the downward transition, your horse should halt with lightness and stand square. The hind legs should be well under the horse, as opposed to sticking out behind him.

Fixing Mistakes in Halt

If the horse steps back at the point of halting or throws his head up, you can usually blame the rider's hands. The problem is

▼ A good halt is square and balanced.

likely that the rider has continued to hold against the horse's mouth, instead of yielding or lightening the hand aids.

There are plenty of other mistakes to be made when coming to a halt. Here are a few of them:

NOT SQUARE. The horse should be standing with his legs under him, parallel in the front and in the back. One hind leg farther behind or a front leg more forward means the horse does not have even weight distribution over all four feet. The rider didn't keep riding the back legs up into the halt.

NOT STRAIGHT. When the horse comes down the centerline, the halt comes with a quick swing of the haunches sideways. Most likely the rider is trying to ride the horse straight with the hands, as opposed to using the leg aids as she should be doing.

ON THE FOREHAND. The horse comes to a halt, but with all of his momentum on his front end. In the extreme instance, the horse can even look as though he's just done a nose dive. Blame the uneven balance between the rider's hand aids and the rider's leg aids for this problem.

TOO ABRUPT. A horse needs to be prepared by the rider for the halt by being given a series of half halts to make the transition to halt smooth and seamless. When a rider just pops the halt on a horse, the reaction can look like that of the driver of a car who just had a deer jump into the road in front of him — slam on the brakes!

TAKES TOO LONG. The opposite of an abrupt halt is the horse who

▲ This halt is not square; the horse's weight is not distributed evenly.

comes to a halt like a decelerating freight train. This horse is either running through the aids by simply ignoring them or is not getting a clear enough message that the rider is after a halt.

NOT STEADY IN THE HALT. This horse hasn't learned that the halt should be held until the rider asks for something else. He throws his head up, moves sideways, and starts to dance around.

GOOD TRAINING HABITS MAKE GOOD HALTS

Perhaps the most common mistake of all is that riders don't include halt work in their daily routine. When there's a busy list of things to work on with a horse — better trots, transitions, working more through the back — the halt often gets left out. At least until right before a show, and then this neglected exercise is suddenly very important.

After all, in dressage competitions, it happens twice in the same test (for combined training the halt is required only once in a test). Is it any wonder that a horse becomes resistant and tense in a test when asked to halt correctly when he hasn't been asked to do so since the last show?

Part of the problem is that during training, we tend to use halt as a time to do things like adjust the girth, take off a jacket, or talk with our instructor. While this is practical, make it a habit to practice a real halt, then walk on a loose rein, then come to another halt casually, so that your horse can have clarity about when he is working and on the aids, even in a halt, and when he is not.

#17 Rein-Back

I F YOU WORK THROUGH THE GAITS and the halt in a methodical way, you will have accomplished all of the elements necessary for your horse to have his full repertoire of movements that are basic to any type of riding except one: the rein-back. Horses *can* go in reverse, and the rein-back is the movement that allows it.

The rein-back is not something you want to introduce early in your horse's training. Many people think a horse should be able to back up with no problem, but a correct rein-back is actually not all that easy, and it's not considered an early training element.

Definition of Rein-Back

When asked to define a rein-back, a nonhorse person might give the simple (but incomplete) answer that the horse is going backward, and the rider is using the reins to ask for the movement. That doesn't give the whole picture. Although a horse in a rein-back looks as though he's just walking backward, the legs are actually moving in a two-beat rhythm that is more like a trot.

Once again, the FEI has a formal definition for us: "Rein-back is a rearward diagonal movement with a two-beat rhythm but without a moment of suspension. Each diagonal pair of legs is raised and returned to the ground alternatively ... the horse should remain 'on the bit,' maintaining its desire to move forward."

Why Is the Rein-Back Important?

Rein-back is considered a sign that your horse is "letting through the aids." That concept is discussed at greater length in the chapter on the training scale (page 81), but the key idea is that your horse is a willing and obedient partner, always ready to answer any question you ask of him. In the overall picture, a rein-back tests your horse's obedience. The rein-back also helps develop a horse's suppleness and improves collection.

How to Ask for Rein-Back

Before you can work on a rein-back, you have to establish a correct halt. The horse needs to be standing square, with his weight evenly distributed over all four of his legs.

The aids for a rein-back are much like the aids for asking your horse to go forward from a halt. Use both seat bones and alternate legs as though you are

▼ A GOOD REIN-BACK

asking for the forward motion of walk. The moment you feel the horse responding and starting to go forward, use a nonyielding rein, which changes that forward energy into backward movement.

Once your horse starts to back up, your response should be to lighten the hand aids while continuing to maintain the contact. In the beginning, a few steps back is a sufficient response. In time, you should be able to ask your horse for a specific number of steps back.

Fixing Mistakes in Rein-Back

In addition to a thorough definition of rein-back, the FEI lists the faults that are often seen when doing a rein-back. "Anticipation or precipitation of the movement, resistance to or evasion of the contact, deviation of the hindquarters from the straight line, spreading or inactive hind legs, and dragging forefeet are serious faults."

When dealing with any of these problems, start slowly in your diagnosis and treatment of the issues. Backing your horse from the ground will give him the confidence to understand that his body can back up.

From there, one solution is to try to take some weight off his back when you are asking for a rein-back by sitting a little more lightly. This doesn't mean lean forward; just lighten your seat bones. Also, make sure you're asking when your horse is warmed up and moving in a loose and relaxed frame. Tension will certainly cause resistance and a deviation of contact.

▲ This rein-back is rushed, tense, and out of balance.

SCHAUKEL

Although this book does not address advanced-level movements, it is important to know that the rein-back ramps up in difficulty as a horse becomes more highly trained. There is an exercise called *Schaukel*, a German word for a combination of two rein-backs with walk steps in between. (The basic rein-back must be firmly established before teaching a horse the Schaukel.)

The reason the Schaukel is so tough on a horse is that the sequence of the horse's feet when walking forward is different from that when he moves backward. Remember, a walk is a four-beat gait and the rein-back is a two-beat gait. As a two-legged creature, we humans have a hard time wrapping our brains around the difficulty that a four-legged horizontal creature must have when the footfalls abruptly switch.

One additional note about Schaukel: it is translated as "seesaw," and, unfortunately, in the English equine vernacular, seesawing has come to mean a dramatic swinging of the horse's nose, head, and neck from side to side. That seesawing is a training and riding technique that is highly (and rightly) criticized as an incorrect method of riding a horse into contact. The common terminology with divergent meanings can cause mountains of confusion.

#18 Transitions

THE ULTIMATE GOAL of these riding fundamentals is to establish your horse as a true athlete, with obedience and rapport between horse and rider. Transitions are a key part of this process. Getting your horse to move smoothly from a canter to a trot, for example, tests the athleticism of your mount. Unless he's properly developed, he'll fall out of the canter into the trot, rather than moving with coordination and balance.

Obedience is obviously a key, too; without it, the transitions will be sloppy or mistimed.

What's a Transition?

A transition is a change from one gait to another or changes/variations within a gait. The FEI offers more detail: "The changes of pace and variations within the paces should be exactly performed.... The transitions within the paces must be clearly defined while maintaining the same rhythm and cadence throughout. The horse should remain light in

Although great riding makes it look as though the horse has ESP, in reality the horse doesn't know what's coming unless the rider tells him.

hand, calm, and maintain a correct position."

You can break down transitions into two categories: transitions between the gaits and transitions within the gait. The more obvious of the two is between the gaits, where you have upward transitions (moving from a lower gait to a higher gait, as in trot to canter) and downward transitions (from higher to lower, as in trot to walk). An example of a transition within a gait is asking for a trot that goes from longer strides to shorter strides.

Why Are Transitions Important?

Transitions are not just gearshifts for your horse; they are training exercises unto themselves. When you make a transition upward from trot to canter, for example, you are working on the athletic training of your horse and his responsiveness to your aids. The same is true for all transitions, whether within a gait or from one gait to another.

If you think of good riders as having a toolbox full of useful exercises, techniques, and ideas for properly developing and training a horse, I assure you that the transition is going to spend very little time sitting around in the toolbox waiting to be used.

Transitions are also essential to the overall training of a horse,

measured by the training scale. In the next several chapters, the training scale will be dissected into its many parts and pieces, but, for now, it's important to understand that transitions make the training scale possible. Without properly executed transitions, a horse will struggle for balance, rhythm, and looseness.

TRANSITION TRIVIA

Here's a trivia question for dressage riders: In any given test, how much time does the judge spend assessing the quality of your transitions?

The answer is somewhere around the 50 percent mark, meaning that roughly half of your dressage test is being judged on how well you and your horse handle transitions. This should tell you just how significant transitions are.

Read the Directives and the Purpose in each test from training level right on up through Grand Prix, and you'll find transitions in abundance. I once heard a long-time judge remark that flying changes didn't impress her, but good transitions did.

What Do Good Transitions Look Like?

The hallmark of a good transition is that a person watching can't really see the communication between horse and rider as the transition happens. Picture the smooth flow of movement between trot and canter or the balanced, seamless energy shift from canter to walk. These are in contrast to the horse who is urged into canter from a speed-demon trot, or the horse whose mouth is taking a beating as his rider tugs and pulls to slam on the brakes to bring him from canter to walk.

How to Ride Good Transitions

Achieving smooth, correct transitions starts with the premise that you are riding with well-organized and coordinated aids: seat, leg, and hands, as always. The horse must be responsive to the leg as a driving aid and accept the contact with the rider's hands.

Well-ridden transitions also require half halts (see page 124). The key is that a horse must be prepared for whatever transition you are asking for. One of the biggest errors a newer rider makes is to assume that just because *she* knows that she is about to ask for a canter, the horse knows, too.

TOP: A rider prepares her horse for the canter with a half halt.

MIDDLE: As the horse responds to the half halt, the rider uses her seat, leg, and hand aids to indicate a left lead canter.

BOTTOM: As the horse jumps into the canter, the rider releases the inside rein to allow the horse to continue in the gait.

That's where half halts come in. With well-timed half halts, a horse will be balanced and tuned in to the rider, waiting for the next bit of information. You can almost see a horse think, "Okay, got the message that something is about to change, and whatever it is, I'm ready."

Fixing Mistakes in Transitions

When a horse is surprised by what the rider is asking of him, his reaction is obvious and predictable. And negative. The horse will typically do one of the following: fall on the forehand, brace and resist, throw his head up, hollow his back, or swing his hindquarters out.

The best way to tackle the problem of bad transitions is to break down your transitions into the smallest parts. Take a walk-to-trot transition for example. First, you need to establish a rhythmic, relaxed, connected walk. When you are getting ready to transition to trot, make sure you have your horse's attention. The correct aids are used to soften the feel just a smidge to say, "Hey, I've opened the door," then give correct leg aids to say, "Trot now." The horse should go right from a four-beat walk to a two-beat trot.

If the horse needed four asks of the leg before responding, threw his head up in the air, or gave any other sign of resistance, come back to the walk and try to clarify your aids again. In the process of coming back to the walk after a bad upward transition, be sure to prepare your horse with half halts so that this downward transition is good. After a few repetitions of these transitions, your horse will get the picture that he should be listening carefully as soon as you give him a half halt. If your aids as a rider are consistent and correctly applied, he'll soon want to answer you the moment you knock on his door.

THINK UPHILL

It's easy to see why one talks about an upward transition from a lower gait to a higher one — after all, you are asking the horse to increase his energy and effort. But the problem is that thinking about downward transitions often results in a horse who simply collapses into the lower gait without energy or smoothness.

One insight into how to make your downward transitions better is always to think "uphill" rather than "downward." For example, if you're cantering and want to make a transition to trot, give your half halts, and once you have your horse's attention, think of riding your horse from canter to trot as though going up a hill. This will keep you tall in the saddle and keep your horse from crashing downward on his front end. This way, he'll transition *up* into trot with the strength of his back end.

▲ His pinned ears and swishing tail show that this horse was not prepared for the transition asked of him.

CHAPTER 5

THE TRAINING SCALE

If there is a Holy Grail in the world of riding, the training scale is it. Also called the training pyramid, scales of training, and training tree, it's a systematic training concept used by horse people around the world.

YOU FIND VARIATIONS in the training scale among countries, with the French having one version and the Germans having a slightly different one. The U.S. training scale is based on the German, with some points of difference. Whatever the variations, trainers all over the world agree that the concept of the scale is to train to the nature of the horse. Permeating every layer of the scale is the moral ground of being humane and fair to the horse while providing the proper building blocks for his athletic development.

To understand the training scale is to understand and appreciate classical riding. Like all of the arts that have the word "classical" attached to them, classical training has stood the test of time. Classical music is the foundation for rock, pop, country, and gospel, and classical riding is the ultimate foundation for jumping, cross-country, and all other horse sports. When a rider understands the ideas and principles of the training scale in a deep and intelligent way, the results for the horse (and the rider) are consistent and reliable.

History of the Training Scale

Scientific theories are an evolutionary process, and the training scale is no different. It wasn't designed and invented in a single day. Training concepts have been evolving since the days of Xenophon, 400 years BCE. As with the world of science, mathematics, and all things artistic, the Dark Ages provided a blackout period for the progress of riding theory. And as with many other parts of our culture, that darkness ended with the beginning of the Renaissance.

Historically, the concepts of the training scale are often associated with various military riding experts who "discovered" aspects of classical training. The German and French military riders were especially important. In the seventeenth century it was a man by the name of Antoine de Pluvinel who recognized that reward instead of punishment is a much more productive way to train.

François Robichon de la Guérinière, in the eighteenth century, invented the dressage seat (the classical position), and realized the effective use of outside aids, among many other riding theories that continue to be used to this day. Alexis-Francois L'Hotte (1825-1904) offered the idea of riding a horse quietly, forward, and straight. German riding master Gustav Steinbrecht (1808-1885) was also an advocate of riding a horse forward and straight (some say in reality too forward).

Over the years, many military riding books were published, edited, and republished. There were disagreements back and forth between the French and the Germans, and although in North America riding is strongly influenced by the Germans, the French contribution to the training scale shouldn't be overlooked.

The Modern Training Scale

By the twentieth century, the ideas that were firmly based in western Europe became the international point of view, so much so that the FEI bases its worldwide rule book for trainers, riders, and judges on the training scale as developed by these men. The training scale is still evolving today, and there always have been and probably always will be endless discussions about the right way to ride and train. The training scale, however, at its core, is the universally accepted theory for proper training.

#19 The Training Scale

THE BEAUTY of the training scale is that it forms the foundation and framework for developing any horse into a happy, well-trained athlete. Although many think of the training scale as only for the competitive dressage horse, it is helpful and effective for every horse: jumpers, eventers, gaited horses, and even those ridden in Western disciplines. While it's primarily considered a process for introducing a young horse to training, it is also correctly used by professional riders up to the highest levels to maintain a horse's physical, emotional, and mental health.

Even at the most casual level, every person interested in horses should understand the basic concepts of the training scale. This includes those taking introductory riding lessons, pony clubbers,

4-H members, and backyard riders. It is amazing how many people who train, show, and even teach never think about the training scale. In my experience, many instructors don't know the training scale other than to acknowledge that it exists, and some don't even know that. This lack of focus and true understanding of riding theory is a profound negative for anyone wishing to ride correctly. Even more disheartening, ignorance of the training scale harms every horse who is subjected to a lack of knowledge and intelligence on the part of the trainer or rider.

Using the Training Scale

It's important to view the training scale not as a method but rather

as a broad framework. All horses are different, and each should be trained and ridden according to its individual needs. What works for one horse in the details might not work for another. But the training scale provides the consistent outline for all.

The rider and trainer also need a sense of time and an understanding of how long a horse should be given to achieve a certain level of training before he's pressed to do more. A talented horse should not be held back, and the less talented should not be pushed beyond what he can physically and mentally handle. Unfortunately, there is a current trend toward not fully developing a horse's basic training before moving him along, especially if that horse is wildly talented. This often ends with

THE TRAINING SCALE

PHASES

ELEMENTS	Familiarization	Development of propulsive power	Development of carrying power
Rhythm	Rhythm		
Looseness	Looseness	Looseness	
Contact	Contact	Contact	
Impulsion		Impulsion	Impulsion
Straightness		Straightness	Straightness
Collection			Collection

disastrous results when the horse is pushed too far and too fast.

In the coming pages, the training scale will be broken down into its parts, with a close-up look at what you're trying to achieve, common mistakes, and how to make your training better. First, here is a brief overview of the elements of the scale and how they connect to one another.

Familiarization Phase

Of the six parts of the training scale, the first section (first three elements) is commonly referred to as the familiarization phase. A wonderful trainer of mine in Germany called this "a time to make friends." In this basic phase, you're looking for your horse to have rhythm and regularity in all three gaits. The loss of rhythm at any level of riding is always considered the most serious of faults.

The next part of this basic phase is a concept that doesn't fully translate into English. The German word is *Losgelassenheit*

(pronounced lows-geh-LAHS-en-hite). The best translation is "looseness," but it's also commonly referred to as suppleness, and in the United States it's often called relaxation. This concept should be like the air that we breathe: it always needs to be there and is irreplaceable if it's gone.

The third element is contact, also often referred to as connection. This refers to the horse going forward to the bridle regardless of whether the horse is in a "dressage frame," jumping, or riding cross-country.

Development of Forward Thrust or Propulsive Power

The next tier of training scale elements makes up the development of propulsive power or forward thrust. It's at this point that a rider is taking natural gaits and turning them into educated gaits. This stage builds on the idea of looseness and contact, then adds in impulsion and straightness.

Impulsion is the development of the energy of the horse through the back. Straightness sounds like a no-brainer, but in reality, straightness always needs attention from the rider, who needs to correct the natural crookedness and difficulties in a horse's body from left to right and right to left. Equate this with being either right- or left-handed, rather than being ambidextrous.

Development of Carrying Capacity

The final section of the training scale groups together impulsion, straightness, and finally collection. Collection can't happen if all the preceding elements are not in place. In collection, the horse is using his hocks and stifle joints to bend more and take increased weight on the hind end. This moves the horse into being more in balance, and he can move with something called self-carriage.

Letting Through the Aids

When all of the parts and pieces of the training scale are achieved and a horse can be ridden in all three gaits without dropping the ball on any element, the horse is said to be "through;" "letting through the aids;" or having, in German, *Durchlässigkeit* (pronounced dorkh-LESS-ig-kite). This is what every rider hopes to achieve when she gets on a horse, whether for dressage, jumping, or simply pleasure.

PROVIDING CLARITY FOR THE HORSE

Horses are incredibly tolerant creatures. They put up with a slew of mixed messages, blurred training methods, and confusing information. Too many horses come to the arena with no clarity about what is expected of them. Their world is a haze of gray because of their riders' vagueness about their sport.

The training scale provides a structure, an outline, a guidepost. A horse and rider who are working within the ideals of the training scale can turn that world of gray into very clear lines of black and white by knowing right from wrong for the effective and positive training of the horse.

#20 Rhythm

▲ Rhythm is the cornerstone of all riding – think of it as "equal distance in equal duration."

"I GOT RHYTHM, I got music ... who can ask for anything more?" George Gershwin made that line unforgettable in the 1930s. When riding a horse, if you've got rhythm, you do have music, and this is where correct training and riding begins.

I can always tell when a riding student excels in music (or math, for that matter), because getting in tune with the proper rhythm of the horse's gait is in great part tapping into an inner talent. Musicians and dancers do tend to have an easier time in the early stages of learning how to ride.

Finding a horse's rhythm is so much like finding a musical rhythm that some riders have been known to school their horses with a metronome to help them mark the correct beats of the gait they are riding. That's not to say that the rider who can't carry the slightest tune will never be in touch with a horse's rhythm, but there is a very clear connection between the two.

The FEI doesn't specifically define rhythm in the organization's rule book, but it is mentioned numerous times. Most often, the word that is married with rhythm is regularity, and here the FEI is very clear: "The regularity of the paces is fundamental to dressage." Consider the correct rhythm of a horse's footfalls to be "equal distance in equal duration." I first heard this phrase years ago in Germany and it has stuck with me since then.

The Importance of Rhythm

Moving rhythmically comes naturally to a horse. From the day a foal is born and learns to romp and run through the fields, he is moving in rhythm. Unless a horse has suffered some injury or has some defect, he will naturally move with rhythm and regularity in his gaits.

In the walk, the horse's feet hit the ground at different times, making it a four-beat gait. Within this four-beat walk there are eight phases. At the trot, the horse is moving two diagonal pairs of legs at the same time, producing a two-beat trot in four phases. In the canter, the horse strikes off and hits the ground at three different times, making the canter a three-beat gait with six very clear phases.

When describing why rhythm is so important, trainers often use the analogy of building a house, with rhythm being the foundation of that house. Building a house on a beach would be a disaster when the first storm comes ashore, unless a mechanically and structurally sound foundation is in place. You can't build safely on the sand; you need something strong as a foundation. The horse's gaits are the foundation for everything else we do in riding.

As with all parts of the training scale, rhythm is not just for the dressage rider. The hunter making a five-stride line between fences is cantering in rhythm, and he's being judged on that steady, consistent rhythm. A horse going cross-country can't

▲ Horses are born with natural rhythm, as this yearling filly shows.

effectively and safely maneuver through a course without consistent rhythm. Imagine a horse being lunged for a vaulting team and not cantering rhythmically. How could a vaulter ever perform the gymnastic movements with a horse whose gait is not reliably rhythmic and regular?

The golden rule in training is that if the rhythm is missing, a rider needs to stay at this point of the training scale until it is consistent, no matter how long it takes. Because there is a firm science as the basis for classical riding theory, we know that a horse who is not moving in rhythm is causing physiological damage to himself. Moving out of rhythm causes uneven wear and tear on joints, tendons, and ligaments, for example, which can cause serious and long-term injury.

AT RIGHT: A horse needs to be in proper rhythm as he approaches a fence in order to be balanced and coordinated at takeoff and landing.

RHYTHM IS NOT TEMPO

For clarity's sake, it should be noted that many people use rhythm and tempo interchangeably when, in fact, they are two different things. "Rhythm" refers to a sequence that is repeated. "Tempo" is the number of beats per minute.

What Causes Loss of Rhythm?

The experienced and well-balanced rider is able to *let* the horse find his natural balance. An interruption in a horse's rhythmic gait almost always results from a rider simply getting on his back. The weight we bring to his back throws off a horse's natural sense of balance. When a horse is off balance, a loss of rhythm is inevitable.

Riders inhibit a horse's natural way of going in a multitude of ways. Start with a rider who has a poor position: think how hard it must be for a horse to stay in balance if 130 or 150 pounds of upright human are skewed too far to one side of the saddle?

Never underestimate what problems a poorly fitting saddle can bring to your horse's natural way of going. When a saddle pinches or blocks the shoulders, for example, the horse is automatically facing an interruption of rhythm because he's no longer moving freely.

The over-strong hand of the rider is probably the biggest fault of all. The rider who restricts the horse's forward movement takes away the neck as a natural balancing device. If the rider manhandles the gaits, she is corrupting, sometimes irrevocably, the horse's rhythm and regularity.

Problems with the Horse

If fixing rider position were the answer to all rhythm problems, it would be almost too easy. A more complicated problem lies with the older, more heavily schooled horses. Here it often takes a bit of detective work to sort out rhythm problems because there are so many layers of issues.

If a horse has been pushed too fast or has struggled with a particular area of his training, there are likely to be rhythm errors. Every time the horse is asked to work in an environment where there has been past tension, he will revisit the tension and show rhythm

▼ OVER-STRONG HAND

the horse is energized but being held in. The horse moves into a prance-like trot that is reminiscent of the upper level dressage movement called "passage." Unfortunately, this delayed trot comes from tension and has nothing to do with correct collection.

Canter Errors

In the canter, the big rhythm problems are a four-beat canter and a cross-canter. In the four-beat canter, the diagonal pair of legs that should be leaving the ground together hit the ground at separate moments, resulting in an extra beat. A cross-canter is when the horse canters on one lead in the front and the opposite lead in the back.

Bridle Lameness

Bridle lameness or rein lameness is the most serious rhythm problem in the trot. The horse moves with one diagonal pair of legs farther forward than the other, which can make him look lame. Typically, the signs of lameness switch sides when the rider changes direction. Veterinarians often have trouble diagnosing the lameness because they can't pinpoint the origin, and the symptoms are inconsistent.

Although there are various degrees of a horse being rein lame, often the symptoms become more intense when a horse is under saddle. Rein lameness is nearly always caused by a rider with an over-powerful hand, making the horse's back muscles tense and stiff, which prevents his hind legs from being able to freely swing forward under his body.

Fixing Rhythm Problems

If a young horse — or even a confirmed, well-trained horse — is out of rhythm when being ridden, the solution most likely lies within the rider's aids. Referring back to the chapters on aids and how to use them, a rider needs to use her seat, legs, and hands to guide and support the horse for a return to rhythmic and regular gaits. Her hands need to be sensitive and educated to guide the horse to a more balanced way of going.

The hands, however, are never used alone. Keeping rhythm and regularity in the gaits needs steady, supportive driving aids. When a horse is ridden forward into correct, elastic contact (more on that in upcoming chapters), a horse will gain longitudinal balance. When a horse is ridden with diagonal aids that support him from side to side, the horse is ridden with lateral balance. A balanced horse is one who can move with correct rhythm and regularity.

#21 Looseness

THE GERMAN WORD *Losgelassenheit* no doubt expresses this concept best, at least in a single word. The English alternatives really don't do it justice. It's a long word, but it's easier than trying to say "looseness, relaxation, calmness, letting go, and suppleness" every time.

The English translation of Losgelassenheit (lows-geh-LAHS-en-hite) can get a bit muddled. It means looseness in the horse's body, suppleness in the horse's body, and relaxation in the overall way of moving. The FEI doesn't issue a specific definition, but looseness is usually described as a physical relaxation of the horse, such that he is working freely through his back while being mentally alert.

The trouble with defining Losgelassenheit is that it's not like the mathematical precision of rhythm. It's hard to quantify because it involves a somewhat subjective assessment of the overall impression of the horse in motion. If riding is both an art and a science, Losgelassenheit moves from the science into the art side of things.

Why Is It Important?

Losgelassenheit is one of the "must-haves" in riding. No work makes sense unless a horse is properly supple. Losgelassenheit is considered a prerequisite for all training because when you have it, all else is possible. Your horse works economically in his muscles, he's flexible in his joints, he's unconstrained in his movements, and he's mentally relaxed.

Without it, your horse is tense, stiff, and not moving properly. Veterinarians everywhere deal with damage to joints, muscles, and tendons inflicted on horses who aren't working with looseness.

Some classical riding descriptions place looseness before rhythm, which indicates just how important it is. Some argue that it's the absolutely central theme that comes before everything else. Others have tried to put it on the same tier as rhythm.

It currently resides one level after rhythm, based on the logic that a horse can move rhythmically without being loose and supple, but he can't move in complete relaxation if he's making rhythm mistakes.

The subject of looseness focuses intensely on a horse working freely through his back. In the anatomy of the horse, the back is like a bridge, and the topline (from poll to tail) needs to act like an arch so that he is able to carry a rider.

The horse's back becomes a bridge when he's stretching his neck forward. When this happens, the area between his withers and sacrum is stabilized. The more the horse is able to stretch forward and bring his hind legs forward, the more weight he can carry.

Shortening the top muscle chain is a huge negative, because it produces a hollow, tight back. Shortening the lower muscle chain is good because the back lifts and becomes a bridge.

▲ The horse on the left is working freely through his back, while the one on the right shows the effects of a short, tight topline.

Signs of Looseness

When you work a horse (or watch a horse being worked), there are a number of visual and audible clues as to the horse's level of looseness. Here is a head-to-tail checklist:

TOPLINE. Start with the overall topline, poll to tail. Is the head up and the neck coiled back against the withers? Is the back down and hollow? These signs indicate a shortening of the topline muscle chain. The neck should be reaching forward, with a back that lifts up as the legs swing under the belly.

HEAD. The head should move gently with the motion, as opposed to shaking or twitching. Look for a soft, relaxed eye. It should appear kind and happy, with no white showing around the edges. A tense eye will remind you of the "deer-in-headlights" look.

The ears should be relaxed and not pinned back. Some horses relax to the point of having nearly floppy ears.

MOUTH. The mouth can be particularly telling. A horse who chews gently on the bit, showing a soft white foam around the mouth, is relaxed. (Not all salivation is a good sign, however. Threadlike saliva indicates stress.)

You should also note the difference between soft chewing and grinding teeth. The horse who goes around the arena grinding his teeth loudly is tense and stressed.

TAIL. A relaxed horse has a tail that moves with the rhythm of the gait. A horse whose tail is swinging back and forth like a fly is driving him nuts is not happy about something. A quick swish when responding to a cue, like a canter transition, is one thing, but when the tail is wringing constantly, it might earn the nickname of "tailicopter."

If the tail is hugged close to the body or tucked between the legs (picture a dog getting in trouble), the horse is showing a lack of looseness through the back.

BREATHING. The relaxing horse often snorts or gives a gentle blow through his nostrils, which indicates that he is chilling out and moving with relaxed muscles. This looseness comes from within — the horse's respiratory muscles relax when the overall body is moving without tension.

Think about when you're nervous or tense and your breath becomes quick and high in your chest. When you calm down, you breathe more gently and deeply.

GAS IS GOOD. Another sign that comes from within is gas that your horse releases from his intestines. When tight and tense, a horse clenches his entire body. After muscular relaxation begins, the natural gas from digestion and intestinal function is released. (This is different from the frequent release of manure, which is often a sign of nervousness.)

▲ The differences between a horse moving with looseness and one without are clear. Compare the mouths, ears, heads, necks, toplines, legs, and tails of these two horses.

How to Achieve Looseness

The methods of encouraging your horse to work freely with a swinging back fall into the category of gymnasticizing your horse. All riding sessions should begin with a solid 10-minute walk warm-up before moving into a rising trot.

If your horse has come in from a paddock, where he's already had a chance to move freely, the walk portion can be shorter. But if he's coming from a stall, the 10-minute warm-up walk is an absolute necessity and should never be bypassed in the interest of "getting down to work."

The horse needs this time to relax into a working mode and stretch his muscles, joints, tendons, and ligaments. A good warm-up in rising trot includes working both directions of the arena on circles, bending lines,

and a limited number of straight lines, all in a slightly longer, freer, relaxed frame. You should focus on circles and bending lines because any time your horse is bending his body, he's naturally becoming more supple.

Following this work, the horse can do some trot-to-canter transitions. Start with a little lighter seat to allow the horse the freedom to warm up his back. Again, you should be working both directions on circles, bending lines, and a few straight lines.

If your entire ride seems like one big warm-up to get to a point of looseness, that might be just what your horse needs. (Remember the idea that no work makes sense unless the horse is loose and supple?) Creating a loose, relaxed, supple horse is a goal as well as a means to a goal.

Some Training Options

When working for Losgelassenheit, success comes from a rider's attentiveness to what achieves results and what doesn't. There is no one recipe; each horse, depending on age, mental state, stabling situation, level of training, and so on, needs something different. When looking at the variety of options, think about the gymnastic effect of each and how to use it most effectively.

Consider using circles that decrease and increase in size, spiral circles that use a leg yield to return to a larger circle, transitions within the gait, leg yields down a long line (e.g., from a quarter line to an outside wall), and gymnastic jumping grids. Here are several other exercises for achieving looseness.

▲ Lungeing is an ideal way to allow your horse to work into a loose, relaxed frame before adding the weight of a rider.

Working at the Lunge

Lungeing is a highly recommended method of working your horse into a relaxed frame. You'll be the most effective if you keep your lunge circles big — 20 meters — and don't work too long. Twenty minutes or so is about the maximum.

When lungeing, consider whether your horse is stretching forward and swinging through his back. Work transitions and aim for more reach of his hind legs under the body. The beauty of lungeing is that you take any negatives that a rider brings to the saddle out of the equation.

Taking Time to Trail Ride

Take a nice hack through the fields before you work in the arena. Many riders do not have this option, but if you do, take advantage of the opportunity. Hacking gives your horse (and you) a breath of fresh air both figuratively and literally.

Imagine if your life was always within the four walls that surround your work. An outdoor environment relaxes a horse and makes him feel freer. The varied terrain has athletic benefits for him as well.

If riding outside of an arena makes you nervous, ask your instructor to take you — or a group class — out once in a while. Both nervous riders and horses benefit from company.

Using Cavalletti

Setting out an arrangement of cavalletti requires a bit of advance planning, but it's well worth it. Cavalletti can be used when lungeing, as well as when riding at walk, trot, and canter. You can set them at varying distances and varying heights and ride them on straight and bending lines. Cavalletti work asks your horse's body for different movement and focus. This can help to free the horse's back.

TOP: Working over cavalletti helps your horse loosen up and swing through his back.

BOTTOM: A hack through fields or woods can offer a relaxing change of pace for both horse and rider.

Testing for Looseness

When all is said and done, how can you judge whether you've developed looseness? The best test is when you let your horse stretch on a circle. This test comes with a variety of names depending on who's teaching: "long and low" (my preference), "chewing the bit out of the hand," "stretchy, chewy," and "stretchy circle," to name a few.

Regardless of the name, the concept is the same: the rider allows the horse to gradually stretch forward and down while moving on a circle. This test can be asked in walk, trot, or canter. Not every horse will look the same doing it, given differences in conformation and level of training. There are, however, some very clear right and wrong methods.

The test is performed correctly if the horse softly moves into the freedom given by the rider's giving elbow and hand. When the horse moves to reach into the bit, he stretches his neck down and his nose goes slightly forward in front of the vertical. The horse should show the greatest possible stretch without losing his balance and falling forward. The horse should also maintain contact with the rider's hand during this test.

Mistakes in Long and Low

The horse should not curl his nose under to avoid contact, and should not snatch the reins from the rider or root at them in a stop-and-start motion. Another big mistake is if the horse goes above the contact and shows no stretch or reach into the giving hand of the rider.

It's important that the horse's rhythm not change when being asked for a long and low. A horse who loses tempo, speeds up, and starts to race when given relaxed contact is not relaxed. Equally important is the way the horse returns to a working frame. He should not show resistance when the rider shortens the reins at the end of the long and low.

▼ CURLING BEHIND

▼ CORRECT LONG AND LOW

◄ Horses should be allowed to work into a long and low frame at several points in any training session for both physical and mental considerations. Give him frequent but brief chances to reach into a stretchy circle. It helps build his topline as his muscles become stronger, and is a good mental break from the mode of work. Equate it with standing up and stretching after working intensely at your desk or computer for a period of time.

#22 Contact

THE DEFINITION OF THIS PART of the training scale is easy to recite, but the words do not begin to describe the difficulty in getting it right. Correct contact is when the rider has a straight line of contact, bit to hand to elbow. The connection between the horse's mouth and the rider's hand is steady, soft, and elastic.

When I first introduce the subject of contact and connection to a student, I often use the analogy that good contact is like a screen door and bad contact is like a storm door. When you shut the storm door, you block the airflow between inside and out. This is what happens if you lock your horse down with restraining aids and he has nowhere to go.

At the other end of the spectrum, if you throw the storm door wide open, you're open to flies and bugs and anything else that wants to go in or out. In riding, this means you've abandoned your rein connection with your horse and he can go where he pleases.

The real solution lies in-between — something like a screen that keeps the bugs out but lets the breeze and fresh air in. The connection through your hands and arms to your horse's mouth should provide information and support while allowing freedom of movement.

▶ A horse and rider galloping on the cross-country course need contact for balance and control.

Contact vs. On the Bit

Many riders (and some trainers) assume that "contact" and being "on the bit" are the same thing. They're in the same category, but having correct contact is not necessarily the same thing as riding a horse in a dressage frame and on the bit. (By frame, we mean the side-on view of the horse along his topline from nose to tail.)

For example, a jumper being ridden to a fence needs to be ridden on contact, not on the bit, with his nose forward and his head up, as he prepares for the upcoming jump.

In the early stages of training a young horse, contact teaches the horse that when the rider releases a bit of pressure on the reins, he should go forward, and when there is pressure on the reins, he should slow down or stop. Furthermore, a young horse learns to turn through the use of an opening rein. In this phase of training, the rider is teaching the horse to move into two reins with equal pressure to help establish balance.

When you look at a dressage horse under saddle, you see the evolution of good contact, which is what trainers mean by a horse being on the bit. "On the bit" means the horse is stepping underneath himself and engaging his hind end, moving his energy from the hind end up through his back, and reaching forward with his neck and head into the bridle. It sounds like a lot, but it really is all the result of good contact.

Good Contact Is Critical

Good contact and connection are a crucial combination for both horse and rider. Developing the ability to ride a horse correctly into contact, and being able to coordinate the aids and maintain that contact, is probably the most important step in a rider's education. Learning to accept correct contact is arguably the most important step in a horse's education as well.

Correct contact is critical because without it the horse is out of balance. When he is out of balance, the entire training scale crumbles. There can be no rhythm and looseness. The horse is unable to develop propulsive power in the form of impulsion. There will be no straightness or collection.

Contact is one of the places where you see how intertwined the parts of the training scale are. Understanding how the horse moves in a scientific sense (see sidebar) is to understand the interdependent nature of the parts of the training scale: correct rhythm, looseness, relaxation and suppleness, and the necessity for correct contact.

Establishing Good Contact

The first question is who's responsible for the contact, you or the horse? If you remember this phrase, you'll always get the answer right: the horse seeks the contact; the rider provides it. Let your horse *come* into the contact. When a rider is *making* the horse come into contact, she has it backward. A horse must step forward toward the bit.

Using the circle of aids correctly is called "riding your horse from back to front," as opposed to from front to back. When a horse is ridden from front to back, the rider uses a heavy, often restraining hand to hold the horse in a frame and then with the other aids drives the horse into the restraining hand. One sarcastic term to describe this technique is the "yank and spank" method: yank your horse in the mouth and spank him into the bit with your whip.

In the correct, classical mode, good contact results from the correct use of driving and restraining aids. Mistakes in contact are always a sign of uneven balance between the rider's leg and hand.

The Importance of the Poll

The evolution from contact to riding a horse on the bit has many prerequisites: good seat, legs, and hands, for starters, and good rhythm and looseness. If you've been riding a horse forward into equal pressure on both reins and your horse is always moving forward toward the bit, your contact is correct.

As you proceed, you should have the feeling that whenever you lengthen the reins, your horse should stay balanced because he will reach forward with his neck into the softer sense of contact. This shows a horse who is becoming stronger in his back and can move between working positions

◄ With good contact, the horse moves through his whole back forward into a soft, elastic connection with the rider's hand.

(working trot, working canter) and a stretching position (see page 94). It's important that in the working position, the poll remains at the highest point.

As the horse (and/or the rider) becomes more educated, you develop a more refined sense of what the poll can and should do. When the horse reaches this stage, he's ready for the rider to do more with the reins and legs than just steer. The hand aids change from steering aids into framing aids, meaning the hands are asking the horse to be in a frame, as opposed to just telling the horse where to go. The horse is now ready to be ridden on the bit.

For a detailed explanation of poll control, see page 118.

THE CIRCLE OF AIDS

Biomechanically, a horse carries a rider's weight by having his body form a bow or a bridge with his back, as described in the previous section on looseness. When the rider is correctly connected to a horse's mouth, her driving aids push the horse forward from his haunches, starting what is called the circle of aids. From there, the horse releases a chain of muscle function that runs from his hindquarters, over his back, to his poll, to his mouth, and finally returns to the rider's hand, which completes the circle and recycles the forward energy.

Fixing Mistakes in Contact

To assess correct contact, you can't just look at the front of the horse. Contact is the forward push *from* the hindquarters *into* the bit, so you need to look at the whole horse to see what is actually going on. To be technically correct, the poll is the highest point, the nose is slightly in front of the vertical (the goal should be about five degrees), and the horse is pushing forward through his whole back.

When a horse is developing the feeling of contact and learning how to be on the bit, there are several common errors, as discussed below.

Bridle Issues

Problems with contact can start with the incorrect adjustment of the bridle. First, make sure that the noseband isn't pinching your horse's face, as shown here. The flash noseband can easily be overtightened. This horse literally needs a little breathing room.

Conversely, the noseband and the flash can be too loose, allowing the horse to open his mouth and avoid the contact.

Above the Bit

Probably the easiest error to spot is when the horse is resisting and is against the rider's hand with his nose above the bit. Sometimes just a minor adjustment is needed, but often being above the bit is a sign of resistance and the horse will not flex at the poll. In this case, the horse uses the muscles on the underside of his neck to fight against the contact and then stiffens his back.

Ideally, the muscles on the top of the neck should be developing, but when a horse is allowed to go above the bit for long periods of time, the muscles on the bottom are the ones that develop, which exacerbates the problem.

Although this is a serious contact problem, it is generally

considered one of the easier to address. There is never a one-size-fits-all fix for training issues with a horse, but remedies, in this case, include lungeing with side-reins and a greater focus on loosening exercises (Losgelassenheit).

Behind the Vertical

The next category of contact issues can be a little harder to see as well as fix. These four problems indicate a horse who's being ridden from front to back with its nose behind the vertical in varying degrees. When the horse's nose does not push forward into the bit and is slightly behind the vertical line, it is known as being behind the vertical.

Although not considered a huge violation of correct contact,

▼ ABOVE THE BIT

▼ BRIDLE ISSUES

working behind the vertical can be a precursor of bad things to come (such as a false bend or getting behind the contact), and is often the first clue that a rider's hand may be too strong, or her driving aids are not strong enough. Sometimes, the rider's restraining aids are too strong, which brings the horse's nose back too far. It can happen in any gait and results in a horse who is behind the vertical. The fix is to use more driving aids and fewer restraining aids.

False Bend

The problem of false bend is more significant and harder to fix. False bend occurs when the flex-ion point is actually behind the poll. To be precise, the poll is con-sidered the first 4 inches (10 cm) back from the area between a horse's ears. In a false bend, it's very clear that the horse's poll is not the highest point and that the horse is being ridden with strong restraining aids.

A false bend sets off a whole chain of events that shuts down the horse's access to the energy coming from the haunches and creates problems for the horse's breathing, proper balance, and, ultimately, his movement. The fix is to back off the restraining aids and allow your horse to move for-ward into a softer, allowing hand.

▼ BEHIND THE VERTICAL

▼ FALSE BEND

Behind the Contact

Among one of the hardest problems to fix is the horse who has been ridden such that he avoids the contact by keeping the nose back; the horse is behind the contact. Often the horse is curling his nose and neck because of previous training issues. A horse who has had bad experiences with contact — typically shown by a habitual false bend — is reluctant to move into contact. Curling behind the contact is his way of avoiding the overly strong hand he's experienced in the past.

This issue is tough because the horse moves forward but refuses to reach and stretch through the topline. Fixing this requires extreme patience and kindness to teach the horse to trust the bit again.

Nose Wagging

Another common error occurs when a rider has uneven contact with her horse's mouth. Sometimes, it's just a minor left and right pull on the rein, but even a little unevenness is too much. You always need to have equal pressure on the horse's mouth. Otherwise, the unevenness causes him to move his head, neck, and body into an unbalanced position.

Unfortunately, some riders routinely train by deliberately "seesawing" the reins. This is a repeated alternate pull on the left and right reins that swings a horse's nose and head back and forth. (This use of the term "seesaw" should not be confused with *Schaukel*, which translates conceptually into English as seesaw. See page 77.)

The result of seesawing is a horse who tracks forward, but with the nose swinging from side

▲ BEHIND THE CONTACT

▼ NOSE WAGGING

to side or nose wagging. Just like uneven rein pressure, seesawing throws off the horse's balance from head to tail. It's also a sign that the horse isn't seeking the contact — as he should. Instead, this maneuver is the rider forcing the horse into contact. It's not a partnership with the horse; it's just the rider working to make it happen.

Leaning on the Bit

Leaning on the bit is another common problem. This is when a horse gets heavy in the hand and bears into the contact. Although we want a connection with the horse that can be described as pressure, we want that pressure to be light and elastic.

Another way to think of it is positive tension on the reins. When trainers use the term "positive tension," they mean the right amount of tension, when you can feel the horse lightly against your hands.

I often ask students how many pounds of pull they feel like they're holding. In my experience, if the answer is 7 or 10 pounds (3 or 4.5 kg), it's too much. That much pull tells me that the horse is leaning on the bit, which trainers label "negative tension."

This is typically a problem that starts with the horse, not the rider. But it's still your job to fix it. A rider needs to use half halts to move the horse off the forehand and be more balanced with energy from behind. This will result in a horse lifting his frame and carrying himself.

▲ LEANING ON THE BIT

ÜBERSTREICHEN

Here is a great test to check your horse's connection and self-carriage, not to mention the fun in saying a fantastic German word: *Überstreichen* (pronounced EW-ber-strike-en), which means "super give with the reins." Überstreichen is an exercise in which the rider releases one or both reins forward for a few strides in either trot or canter. If your horse is able to maintain the same balance, rhythm, and topline, it's a sign that he's accepting the connecting aids. Successful Überstreichen is considered proof of self-carriage.

In lower level horses, Überstreichen is most often asked for on a circle. Upper level horses can handle Überstreichen in the advanced movements such as collected canter, and even canter pirouettes.

#23 Impulsion

REAL, HONEST-TO-GOODNESS impulsion can be a bit deceiving. The issue is how to differentiate between actual impulsion in a horse and a naturally good mover with large, leggy athletic strides. You can have a horse with both, but it's important to recognize the distinction between the two.

A horse who has been trained correctly with the foundation of rhythm, looseness, and contact will move through his swinging, relaxed back and show a clear thrust from the hindquarters.

Riding a horse with impulsion is something like catching a wave on a surfboard — his back rises up underneath you and propels you forward. Then, you sink down for a moment as the horse prepares to thrust forward again.

▼ Impulsion is only developed at the trot and canter, where there is a moment of suspension in each stride as the horse moves across the ground.

What Is Impulsion?

When a horse starts to use his full body to move and in doing so shifts his center of gravity more toward the center of his body, as opposed to traveling with the majority of his weight on his forehand (front legs), it is impulsion. To gain a deeper understanding of exactly what this means, let's look again to the Germans, who define impulsion as *Schwung* or momentum.

There is a descriptive sound to the word Schwung, even if you don't speak a word of German, and indeed, it can be translated as "swing." Other key English words that describe impulsion are "thrust" and "propulsive power."

The official FEI definition of impulsion is "the transmission of an eager and energetic, yet controlled, propulsive energy generated from the hindquarters into the athletic movement of the horse. Its ultimate expression can be shown only through the horse's soft and swinging back guided by a gentle contact with the athlete's hand."

No Impulsion at the Walk

Impulsion can exist only in the trot and canter; not in the walk. Only the trot and canter have a moment of suspension (when all four legs are off the ground at the same time). Technically, the walk has no moment of suspension and, therefore, no impulsion. The walk still needs its own sense of propulsive, positive energy, but it's best if you think of the walk as having engagement as opposed to impulsion. (For more about the illusion of impulsion in the walk, see page 53.)

To have impulsion, the horse needs to use his engine (hind end) and allow that energy to swing through his whole body into forward movement, marked by a relaxed, free-moving back. A prime indicator of impulsion is the amount of time the horse spends in the air; impulsion is often described as "long in the air, short on the ground."

Another indicator of correct impulsion is that the horse covers more ground in the extensions and has more elevation in collection.

▼ Impulsion is as important in jumping as it is in dressage — look at this horse's back legs as he gathers himself and then releases that energy forward over the jump.

How to Make Impulsion Happen

Impulsion is an athletic development; it's not something where you can just go into the arena, snap your fingers, and *poof,* suddenly it's there. As dictated by the training scale, you need to establish good rhythm, suppleness, and contact before you can develop impulsion. The horse must willingly reach into the bit, and he must be responsive to your aids. In a nutshell, you must have a horse who will accept your hand on the mouth *and* your driving aids.

It's best to develop your horse's impulsion beginning with the trot. Half halts (page 124) are imperative at this stage. The hind end is the key to building impulsion. A half halt makes a horse step more under himself with his hind legs, so no half halt, no impulsion. The quality of the half halt will directly affect the horse's ability to build impulsion.

Using Transitions

A rider begins to develop impulsion in her horse by first lengthening his strides in the trot and then using frequent, well-timed half halts to make a transition back to a working trot. These transitions within the gait can be broken up with transitions between the gaits. For example, use a sequence of working trot to trot lengthenings, back to working trot, followed by a few trot-to-walk-to-trot transitions.

Trot-canter-trot transitions are especially good for developing impulsion. These transitions get right to the core of the horse's anatomical makeup. All the joints of the hindquarters are involved in building impulsion, and riding these sequences of transitions builds the "forward-upward" ability of the horse.

One word of caution about these transitions: if they are abrupt and on the forehand, they are not doing you any good. Think about driving forward *into* the trot in the downward transitions instead of stopping the canter. In other words, you are beginning the trot, not ending the canter.

Cavalletti Work and Jumping

You can also build impulsion with cavalletti work and jumping. Laying out a sequence of low cavalletti can help a horse work on shorter and then longer steps and strides. Setting out a grid for jumping is also excellent. A five-jump line that includes a bounce followed by a pair of one-stride jumps is a fun and effective way to build propulsive power.

▲ You build impulsion by using transitions within the trot or canter, moving from shorter working strides (left) to a lengthening of stride (right) and back down.

Hill Work

Another fun and highly productive training exercise is hill work. Riding a horse uphill in trot and canter builds up his hind leg strength and increases the articulation of the hind joints. There are longer moments of support as the legs move more under the horse and, consequently, the horse's neck and head need to balance the whole body by moving more forward and downward, all of which is great training for impulsion.

Mistakes in Impulsion

Probably the most common mistake is to confuse impulsion with speed. The key is getting energy from the articulation of the hind-end joints and not the actual rate of forward motion. Making your horse run around the arena faster is not impulsion.

Tense, hovering steps is another serious mistake. These trots look as though there is a slight delay before the horse puts his feet on the ground. With this kind of trot, the horse moves with uncoordinated legs and a lack of suppleness. Even though these trots can be showy, they cause incredible wear on the horse's legs and back because of the overall tension in the muscular system.

A BURST OF ENERGY

I like to use the image of a bottle of soda to illustrate impulsion. If you put your thumb over the top (as we all did when we were kids) and shake the soda hard, that's like the collecting phase for the horse: lots of bottled-up energy that's ready to burst. When you take your thumb off the top of the bottle, the soda spurts out like Old Faithful, which is like the horse's energy bursting forth in a medium or extended trot.

▼ HILL WORK

CLASSICAL DEBATE

The subject of impulsion is the cornerstone of a current debate between classical and modern methods of training. Proponents of the classical method argue, convincingly, that the age-old training methods are based on the nature of the horse and work to support the overall health of the horse's body and mind. Defenders of the newer methods say there are considerable advantages and the results (as in high scores at dressage competitions) should speak for themselves.

In German, *Rückengänger* (back mover) applies to a horse with positive energy; active hind legs; supple, swinging back; ground cover; and coordinated limbs. This is opposed to *Schenkelgänger* (leg mover) – a horse with negative energy who lacks suppleness and coordination.

Look at the Back

In the current world of competitive dressage, some people mistake the flashy leg mover as showing great positive energy, to the point that some buyers overpay for these horses. More discerning riders jokingly refer to such a purchase as "the horse with the expensive trot." It's easy to be impressed by the exaggerated glitz of a leg mover. The huge trot with the front legs striking out at a near-90-degree angle can be a stunning thing to watch.

But these front legs often contrast with what the back legs are doing, which is nearly nothing. I once got some excellent advice from a German trainer: "If you want to see how a horse is really moving, don't look at the front of the horse, look at the back."

Unfortunately, judges reward the expensive trot with high marks. But although these horses may wear the title of champion, advocates of the classical approach point out that exceptionally gifted front-end movers are displacing classically trained horses. If the leg movers win while spectators cheer them on, there is a trickle-down effect.

You Don't Always Get What You Pay For

Enamored fans now want their horses to move with the forelegs striking out in front and just forget what the back end is doing. Trainers and riders around the world start looking for and training the big, leggy movers, and riders just getting involved

in competitive dressage, unfortunately, view this as the ultimate trot. But is it good for the horse?

In the foreword to the newest edition of *The Rider Forms the Horse*, first published in 1939, notable German rider and trainer Klaus Balkenhol, recently the USA's *chef d'equipe* for dressage, writes, "Untold numbers of top horses disappear never to be seen again as a result of incorrect training." He continues that, regardless of the equine activity, it is essential that trainers understand the link between muscular function and the skeleton of the horse as the horse works through the different phases of training and exercises.

For our purposes in understanding and developing impulsion, we simply need to remember that impulsion is the full body of the horse at work and, in particular, the hind end, not just the legs.

▼ LEG MOVER

▼ BACK MOVER

#24 Straightness

ON THE FACE OF IT, you would expect that straightness would be among the easier things for a horse to have. Just put one foot in front of the other and you've got straightness, right? Even the basic definition makes it sound easy: straightness is when a horse tracks with his forehand in line with the hindquarters on both straight and bending lines and in all three gaits.

For example, you want to have your horse straight on the circle. While it sounds like an oxymoron, it really means that the horse's feet — front and back — straddle the line of the circle, which is straight in riding terms.

The definition sounds like a statement of the obvious, and I'm often asked why straightness is so deep into the training scale since new riders assume it's an easy task. You might think that it's like a horse being born with a natural sense of rhythm, but the opposite is true. Horses, as humans, are born naturally *not* straight.

The Lack of Straightness

To understand how this natural lack of straightness is possible, let's first examine the actual construction of a horse. It's common to think of a horse as shaped like a four-wheeled vehicle: symmetrical over an even wheelbase, with front tires directly in line with back tires. But, in fact, horses are built more like triangles, with a narrower front end and a wider back end. Have you ever tried to lead a horse out of a stall without opening the door the entire way? His shoulders made it through, but his hipbones probably bumped the doorframe.

Now add in left-brain or right-brain dominance. Just as humans are right-handed or left-handed, horses are naturally one-sided. The dominant side creates stronger muscles and the non-dominant side is looser, creating a natural imbalance in the same way that a right-handed person tends to have a stronger right arm.

This lack of straightness is called "natural crookedness." Just as most people are right-handed, the vast majority (between 80 and 90 percent) of horses are naturally crooked to the right, meaning that when they track right, they move in a crooked manner. When a horse is crooked to the right, we say his body is stiff on the left and hollow on the right.

Several Kinds of Balance

There is an unbreakable link between straightness and balance. Balance is a tricky word and concept because there are so many variations, as described below:

NATURAL BALANCE is what a horse is born with and how he moves at liberty. Like people, some horses have better natural balance than others.

ADJUSTED BALANCE is how the horse takes into account the weight and movement of a saddle and rider.

LATERAL BALANCE is how the horse handles his weight as well as a rider's from side to side. For straightness, lateral balance is the focus.

LONGITUDINAL BALANCE is from front to back and from back to front, which is also affected by carrying a rider. Longitudinal balance is also an essential part of collection, which we'll get to in the next chapter.

▼ Horses are not symmetrical.

Why Is Straightness Important?

Training a horse to move straight is at the core of turning his natural gaits into trained gaits. In the big picture, a horse can never reach his full capacity of propulsive and carrying power if he isn't moving straight. If we allow a horse to continuously go crooked, he will incur significant wear and tear on one side of his body because he isn't using it evenly.

In a serious case, he will actually suffer injuries over the course of time. Imagine a professional gymnast who only trained on her left side. Her gymnastic potential would never be reached and she would sustain injuries from the right-side weakness. You need to think of equine athletes in the same way.

Straightness is a relative concept. The more correct and thorough your riding is within the context of the training scale, the more your horse can track straight, which, in turn, improves rhythm, helps looseness, evens out contact, and creates better impulsion.

How to Improve Straightness

Before you gear up to focus on riding your horse straight, it's important to recognize that, without the other elements of the training scale significantly in place (rhythm, looseness, contact, and impulsion), a horse will find it very hard to travel straight. The horse must be quick to accept the seat, leg, and rein aids. Even if you are working at a very basic level, you can still mix elements into your daily rides that will help your horse develop better straightness.

Use Bending Lines

Since the overriding goal when working to improve straightness is to balance the muscles on the left and right sides of the horse's body, good training includes bending lines and other work that requires your horse to bend through the rib cage. Take note that while the term "bend through the rib cage" is common in training, it's really the muscles that bend and stretch, not the bones themselves. The skeleton can't actually bend. (See page 118–123 on flexion and bend.)

When riding bending lines, pay attention to whether your horse is falling toward the wall with his outside shoulder or falling in with his inside shoulder as you move off the track into the bend. Also take particular care about where your horse is

---

THE SCIENCE BEHIND STRAIGHTNESS

Picture a horse who's tracking to the right and moving with his right hind foot inside the track of the right front foot. Biomechanically, he is not using his right hind leg very much. He's overusing his left hind leg and working harder on his left front leg, too. When the horse tracks to the left, the dynamics switch and he feels stiff in that direction because he lacks bend and flexion on that side of his body.

The rider feels this problem as an imbalance in her seat and unevenness in the reins. When the horse tracks to the right, he feels heavier in the left rein and has looser contact in the right rein, rather than having equal contact on both reins. Achieving equal contact requires a horse to be correctly balanced, and that means he must have muscles evenly developed on both sides of his body, which, in turn, requires straightness.

▼ Horses are naturally crooked, as seen on the left, and must be trained into straightness, as seen on the right.

putting his back legs in relation to the front legs. They should be tracking in line with the front, not forming a new track. Think of it as riding on a railroad track. Even as that track bends, the horse must stay on the track with his shoulders and all four legs or he'll be derailed.

Work Away from the Wall

In your daily work, develop the simple habit of riding off the wall more than on it. This requires you to be thoughtful with your outside aids, as most horses appear to have a magnetic force pulling them back to the wall. Leg yields can be very useful for straightness, especially if you have your horse move off the wall and toward the quarter line. A counter canter is a more advanced movement, but it's very helpful for straightening a horse.

With all of these suggestions, always try to think of straightening your horse with your leg, not just your reins. Merely flexing the neck is the wrong way to use the reins. Your real goal is to get your horse more evenly into both reins by riding the body into straightness.

LATERAL TERMINOLOGY

For the purposes of this chapter we'll dive a little deeper into dressage theory with a preview of lateral work. Shoulder-in is considered a straightening exercise, and involves riding your horse's shoulders to the inside at a 30-degree angle to the track. The hind legs stay on the main track.

Shoulder-fore, often thought of as a baby shoulder-in, is when your horse's shoulders are moved to the inside at a 15-degree angle. It's also a straightening exercise.

Advanced Straightness

At a more advanced level, achieving the desired straightness involves riding your horse's inside shoulder in front of his inside hip. With this more complicated idea, it's important that we ride the forehand in line with the hindquarters and not the other way around. Remember that impulsion comes from the rear engine; so does straightness.

This concept can be confusing because different trainers call it different things, but all mean the same thing. The common terms are "position-in," "in position," and "first position." Riding in this manner (inside shoulder in front of horse's inside hip) basically means riding your horse with just the slightest hint of shoulder-in. (See box below for more detail.)

By riding in the slightest hint of shoulder-in, you are asking your horse for just a bit of engagement of his inside hind leg. It is easiest to begin this work in the walk and trot. Canter is more difficult because your horse is always slightly bent at this gait.

▲ Tracking crooked with haunches falling in

▲ Inside shoulder lined up with inside hip

#25 Collection

COLLECTION can be thought of as directing your horse's forward motion into upward motion, a bit like a motorcycle popping a wheelie. The FEI definition states: "The aim of collection in the horse is to further develop and improve the balance and equilibrium of the horse.... To develop and increase the horse's ability to lower and engage the hindquarters for the benefit of lightness and mobility of its forehand. To add to the ease and carriage of the horse and to make it pleasurable to ride."

The following gem is among the quotes I have picked up from trainers: "The aim of riding is submissiveness and the aim of dressage is collection." You could say that all roads in dressage and the training scale lead to (or should lead to) collection.

The more a horse develops his carrying capacity and is able to balance himself, the better he is able to carry himself and his rider.

▼ Jumping offers the most extreme moment of collection as the hindquarters lower and engage, showing the elastic movement of the hip, stifle, and hock.

What Collection Looks Like

When a horse is in a collected moment or movement, he's taking more weight on his hind legs and showing more elastic movement in his big joints, which are his hip, stifle, and hock. As he shifts his weight and balance, his center of gravity moves backward as his hind legs step farther forward.

The previous step on the training scale, straightness, is mainly a focus on lateral balance (from side to side); collection is a focus on longitudinal balance (from back to front). Collection is the coiling of the spring, while impulsion is the launching of the spring.

Young horses, poorly trained horses, or poorly ridden horses are often ridden on the forehand, with the majority of the weight on the front end, pulling the horse out of longitudinal balance. Your goal should be an even weight distribution between front and back.

This ideal is never perfectly reached, but when a horse is balanced longitudinally, he is able to move his center of gravity more toward the center of his body, thereby allowing him to better use his back and haunches.

A horse in collection is described as being in self-carriage, meaning that he is carrying himself with his head and neck raised in direct proportion to the degree of collection. The more correctly schooled the horse, the more weight he can carry on his haunches, and the higher he can carry his neck and head (without the rider supporting them) while remaining on the bit.

▲ Collection is the ultimate proof that a horse is moving correctly and energetically from back to front.

DEFINING CADENCE

Hang around a dressage barn long enough, and you're bound to come across the word cadence. Dictionary definitions include "balanced, rhythmic flow, as of poetry or oratory" and "the measure or beat of movement, as in dancing or marching." Both of these apply to cadence as it is used in riding.

In the riding world, cadence means that the moment of suspension when the horse is moving in a collected tempo is clearly noticeable. The USDF Glossary of Judging Terms says, "The marked accentuation of the rhythm and (musical) beat that is a result of a steady and suitable tempo harmonizing with a springy impulsion."

Cadence is introduced in the impulsion portion of the training scale. It's the link between correct, positive impulsion and collection. A horse showing cadence is in the process of transferring propulsive power into carrying power. When you look at a horse who's showing a gait (must be trot or canter) with cadence, you see merely covering ground being transformed into more elevated steps or strides.

How to Get Good Collection

Building your horse's ability to collect or trying to improve his collection means doing exercises that ask him to carry more weight on his hindquarters. As you consider the number of exercises available, be aware that there is a difference between a *collecting* exercise and a *collected* exercise. An example of a collecting exercise is a half halt, while a collected exercise is a canter pirouette.

In the building-strength department, hill work can't be beat. As your horse is gaining strength for impulsion by going up the hills, he's gaining power for collection by going down. Cavalletti work and jumping exercises will also improve your horse's strength.

In your daily arena work, the list of good exercises includes half halts, full halts, transitions between the gaits, circles, voltes, and rein-back. When riding the transitions, aim to really tune your horse in to your weight aids so that your hands don't become the overpowering message center.

TOP: Working on a downhill slope builds strength in the hind joints.

CENTER: Rein-back asks your horse to engage his hip, stifle, and hock.

BOTTOM: Smaller circles and voltes require more activity by the hind legs as the horse reaches underneath himself.

Mistakes in Collection

Difficulties in collection usually relate to the earlier steps in the training scale. Rhythm, looseness, contact, impulsion, and straightness must be firmly in place, and the considerations for each point need to be deeply embedded into daily training before collection can be achieved. Collection also requires strength of the horse, and building strength takes time. A four-year-old is just not going to be physically capable of transferring his balance from front to back.

Collection takes strength, and building strength takes time.

▲ A horse with collection problems tends to hollow his back and brace the neck and head. The back legs are often stuck out behind even as the front legs show elevation and activity.

DEFINING ENGAGEMENT

In the riding sense, the term "engagement" is based on the mechanics of the horse's back end. Using the spring analogy, engagement is when you ask your horse to coil like a spring as in "carrying power." The medium and extended gaits are examples of the "pushing power" or release of the spring. In the FEI rules for dressage, one of the objectives is "the lightness of the forehand and the engagement of the hindquarters, originating from a lively impulsion."

In biomechanical terms, when we are looking for engagement, we are looking for a horse to show more flexion in his lumbrosacral joint as well as in the joints of the hind leg. At the highest level of dressage, this is seen in the weight-bearing phase of the horse's stride when the croup goes lower in relation to the forehand.

The lightening of the forehand allows more freedom in the horse's shoulder, and this leads to more expressive gaits, as can be seen in high-quality medium and extended work in trot and canter.

▲ Using half halts helps a lower-level horse become more engaged and develops a more expressive trot.

▲ In engagement, the croup is lower, freeing the shoulders to move up and forward.

"Expressive gaits" refers to the artistic quality of a gait. It's a feeling or image that has a melding of impulsion, harmony, and cadence. To get the good mediums and extensions, you need to prep your horse by riding him "up more from behind" with the use of half halts.

In jumping, engagement is an imperative. As the horse prepares for a takeoff, he lowers his hindquarters, which frees his front legs and shoulders to lift for the jump. Engagement of the jumper means more clear jumping rounds.

There are two types of movement that are often misinterpreted as engagement. When you look at the hindquarters, it's not just hock action or flexing of the hock joint that counts (some gaited horses move with extreme hock action, or think of the back legs of a Hackney Pony pulling a cart). Engagement is also not the length or reach of a horse's back leg under his belly. Yes, we'd like a horse to move with more reach of the hind legs, but this is different from engagement. True engagement is the actual *lowering* of the croup and *lifting* of the forehand.

#26 Letting Through the Aids

THE TRAINING SCALE culminates with the idea that a horse trained with patience and thoughtful adherence to the classical principles will be a relaxed, obedient, and willing partner. This concept has several names, as seems to be the confusing tradition in riding: "letting through the aids," "letting the aids through," having your horse "through," or, in German, the quality of *Durchlässigkeit* (dorkh-LESS-ig-kite).

Durchlässigkeit and "submission" are often used interchangeably, but the English word isn't exactly right; the concept describes a relationship more in the nature of a partnership. The term refers to a horse's immediate willingness to obey the rider's barely visible aids without the slightest resistance.

The horse is *letting* the driving aids come *through* the hindquarters and recycling that energy back through the rider's hands. The horse also needs to show looseness. Mental or physical tension destroys the picture and the feeling of Durchlässigkeit.

A horse does not have to be at the Grand Prix to prove that he's "through." As any horse progresses in his training, letting through refers to a horse's development of impulsion or *Schwung* (see page 103). At its core, Durchlässigkeit gives evidence that a horse's training thus far has been correct: that is, he is fulfilling all aspects of the training scale according to his skill level.

I'd therefore argue that even a training level horse, who hasn't developed impulsion yet, can still possess Durchlässigkeit, at least at the most basic level.

Improving Throughness

You can improve your horse's throughness by staying focused on the quality of your gymnastic work. On a daily basis, include variations of lines (straight, bending, circles) and directions in your riding. Within these variations, your horse should have many changes between the basic gaits.

All too often, riders become stuck on the sequence of walk, trot, and canter. Mix it up. Have your horse warm up in all three gaits, and then work between the gaits with thoughtful transitions. Within each gait, offer your horse variations in tempo. The rhythm and purity of the gait you're in should remain constant, but the tempo can and should change frequently.

Vary your demands on balance, both lateral and longitudinal, to strengthen your horse physically and mentally. Take the time to consider the strong points that your horse has, and then use these strong points to improve his weak points. For example, if your horse is exceptionally good at collection, you can leverage those good collections into improvements in extensions by working between collection and extensions.

▲ Letting through the aids proves that a horse's training is progressing as it should.

CHAPTER 6

KEY RIDING TECHNIQUES

Technique is an interesting word. The dictionary defines technique as a skillful or efficient way of doing or achieving something; especially the execution of an artistic work or a scientific procedure. Considering that riding is part science and part art, the word is very appropriate.

Some things we do when we ride clearly fall into the category of technique, since they are methods of working with our horse to get the artistic effect we're seeking. The three techniques discussed in this chapter are flexion, bending, and the half halt. In brief, these techniques are both methods to get results in riding *and* components of dressage movements.

Because riding as a sport integrates both art and science, understanding these techniques means knowing both how to do them — the mechanics — and where they will take you — the goal. Position, aids, gaits, and the training scale are the fundamentals of riding, and the techniques are both tools to attain these goals and components of the goals themselves.

For example, flexion is a softening of the horse's poll that helps you keep your horse in balance, so it's a tool for better riding. But, as with bending, it's also a component of many correct riding movements, so it's a goal as well. The same goes for half halts. As a method by which to get your horse's attention and improve his balance, a half halt is a tool in your toolbox. But it's also part of maintaining the right rhythm as you approach a jump, so in that sense it's also a goal.

Building Techniques from the Ground Up

Think of the techniques as being like the nails used in building a house. You might not think of the nails as a part of the house on par with the floor, walls, or roof, but they are an absolute necessity in making the floor, walls, and roof function properly. Techniques in riding are the same. As with the nails in the roof, we don't usually think about them separately, but they're an essential element just the same.

To carry the analogy a little bit further, it's hard to build a proper house without knowing about nails in all their varieties. I don't know the difference between a roofing nail and a flooring nail, but no carpenter would use one where the other is needed.

Riding techniques are the same. You'll spend a very long time trying to get a good, balanced horse when riding a serpentine unless you understand and properly use flexion and bend (and know the difference between the two) and properly execute your half halts.

#27 Flexion

MANY RIDERS USE the terms "flexion" and "bending" interchangeably and why not, since so many other words in the horse world are interchangeable? But flexion and bend are entirely different things. Flexion refers to the slight flexing of the neck and head to the left or right. Bending is a lateral curve through the horse's whole body from poll to tail.

In driving terms, it's like the difference between turning your head at the neck to look at your side-view mirrors and turning the whole length of your spine to look through your rear window.

A horse is flexed when his head is turned slightly sideways at the poll. Sometimes trainers will use the word "positioned" when they are talking about flexion. This means that when a horse is positioned to the right, he is slightly flexed to the right at the poll, and, obviously, if he's positioned to the left, he's slightly flexed to the left at the poll.

Poll Control

Flexion is a study in poll control. Many trainers use the phrase "have your horse flex at the poll." To be clear, a horse can't really flex at the poll because it is the top part of the skull and not a joint.

There are two joints right behind the poll that can flex, however. These joints can be thought of as having a "yes" function — moving the head up and down — and a "no" function — moving the head from side to side.

Anatomically speaking, the *yes* function is known as "longitudinal flexion," as shown in picture A. It happens at the atlanto-occipital joint, where the first cervical vertebra attaches to the skull. The *no* function, or "lateral flexion," occurs in the atlanto-axial joint at the second cervical vertebra.

To have a horse properly flexed or positioned, a rider first has to have the horse soften and give in the first joint (*yes*) and then in the second (*no*). The intended consequence of this movement is that the horse's neck muscles relax and the crest of his neck will "flip" slightly, as shown in pictures B and C. The process also stretches the nuchal ligament in the neck, allowing the entire back to be more elastic.

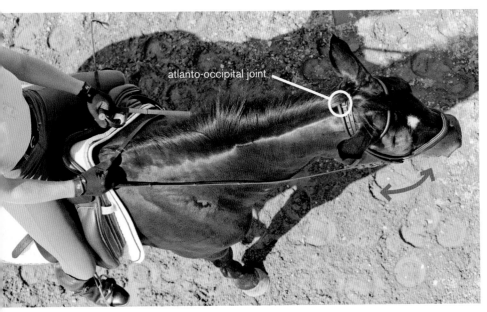

▲ A. The *yes* function occurs at the atlanto-occipital joint.

▲ B. The *no* function occurs at the atlanto-axial joint.

Why Flexion Is Important

A rider wants her horse flexed because this opens the door to having the horse fully responsive to the rein aids. Without flexion, a horse cannot move into the outside rein and cannot accept and yield (soften) from the inside rein — he cannot be properly suppled. Unless a horse is supple, he can't swing through his hindquarters, and without that element in place, the trickle-down effect is that he can't move straight. Note how the training scale disintegrates: no looseness, impulsion, or straightness without flexion.

▲ COUNTER-FLEXION

▲ HEAD TILTING

How to Achieve Flexion

Getting your horse to flex and look slightly in the direction you're going (to counter-flex is to look the opposite way) requires, first and foremost, a balanced rider with good contact to the horse's mouth. Without losing that contact, use a yielding rein on the outside rein and an asking rein in the same amount on the inside.

If the horse flexes correctly, you should feel the horse soften ever so slightly on the inside rein. You should be able to see just the corner of the horse's eye.

Fixing Mistakes in Flexion

A common mistake is asking for too much flexion by overusing the inside rein. As soon as you can see a lot of the eye and more of the face, you've gone too far. This error almost always causes your horse to fall through the outside shoulder.

If the problem continues, it eventually turns into pulling on your horse's mouth, which is a serious issue. The restriction of the horse's neck shuts his forward movement down and physically prevents him from being able to

step forward and under with his inside leg.

The opposite problem is when your horse is counter-flexed. You need to keep inside flexion; meaning, when you are tracking right, you see the right eye, and when you are tracking left, you see the left eye. Losing the correct flexion of your horse is an early sign of a balance problem.

Another issue is when the horse's head tilts to the side, generally because the rider is not yielding the outside rein enough to allow the flexion. The horse ends up carrying his head with one ear higher than the other. Fixing this problem requires the rider to ride the horse straight with no flexion and make him stretch forward and downward (see Long and Low on page 94).

▲ C. The release at this joint allows the crest to "flip."

#28 Bending

BENDING HAPPENS to a horse's body through the entire length of the spine, not just at the neck. With bending, you also have a degree of flexion. An important distinction needs to be highlighted here: you can have flexion without bend, but you can't have bend without flexion.

Why Is Bending Important?

When a horse learns to bend his body through turns, whether they are corners, circles, shallow loops, or any schooling figure, he begins to find that he can stretch and be more useful in his body. This bending creates a more supple horse (looseness on the training scale).

Bending is also an essential element of being able to ride a horse on diagonal aids and vice versa. When a horse develops the suppleness and then the balance of being on diagonal aids, he's on his way to better straightness.

▲ This horse is bending correctly around the rider's inside leg.

There are varying degrees of bend, based on anatomical issues and the level of training that the horse has experienced. Structurally, the horse's body cannot actually bend in equal degrees from head to tail because of limitations at the joints between the vertebrae.

According to studies done by Dr. Hilary Clayton, who holds the Mary Anne McPhail Dressage Chair in Equine Sports Medicine at Michigan State University, the most flexible part of the horse is the area in front of the withers, the area containing the cervical vertebrae. As soon as you get into the withers region (the area called the "thoracic vertebrae"), the vertebrae joints are immobile because the spine in this area needs stability for the attachment of the forelimb. The lumbar vertebrae have some mobility, but the sacrum is completely rigid because the sacral vertebrae are fused together.

Note that, although the thoracic region is without much mobility, there can be a *feeling* of bending for the rider because of the rotation of the rib cage (in the case of an inside leg asking a horse to yield from that leg). Dr. Clayton also has found that the relative motion of the left and right scapulae sliding across the chest wall can give the impression to the rider that the horse is bending.

In regard to the horse's level of training, a young horse who is just starting his work in the arena will not be able to bend as much as an advanced horse. The curve in your horse's body should always match the curve of the bending line. For example, a 20-meter circle has a very soft, gradual bend and a 10-meter circle requires considerably more effort. Obviously, a young horse or a horse just starting his training will only be able to handle the gymnastic work of a 20-meter circle.

DEGREES OF BENDING FROM POLL TO TAIL

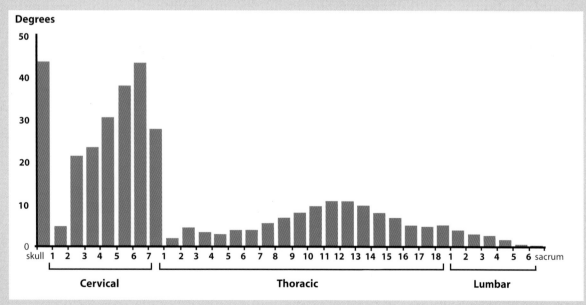

How to Get Effective Bending

The two most important things about teaching a horse to bend correctly are first, that he must accept the inside rein without leaning on it, and second, he must have learned to go into the outside rein. The bend is a balancing act, literally and figuratively, with finely tuned and coordinated diagonal aids, as described here:

THE FIRST AID is to weight your inside seat bone.

AT THE SAME TIME, put your inside leg at the girth line to ask for engagement of the inside hind leg.

YOUR OUTSIDE LEG is in a guarding or regulating position, slightly behind the girth. This keeps his hindquarters from swinging out.

THE INSIDE REIN is asking the horse to be flexed and soft. For younger horses or horses who are particularly stiff and unwilling to bend, you can also use the inside rein as a sideways opening rein to help lead the horse into the turn.

THE OUTSIDE REIN needs to be elastic enough to allow your horse to have correct flexion and bend on the inside, but steady and supporting enough that he can't turn his head too much or fall out through the shoulder.

Asking a horse to bend around your inside leg on his horizontal axis also requires a review of the fundamentals of straightness (see page 107). Horses by nature are quite different left from right. The majority of horses are *stiff* when tracking to the left and *hollow* on the right. The different sides of the horse will present you with different problems.

Fixing Mistakes in Bending

A common mistake is when a rider asks for too much bend in the head and neck. Since the most flexible part of the horse is the area in front of the withers, the neck and head are the easiest part

▼ A stiff horse falls against the rider's inside leg.

▼ This horse is overbent through the head and neck.

to bend and consequently over-bend. You need to focus on getting your horse to bend through the *ribs* around your inside leg. It's bend through the entire body, not flexion in the poll, that you're trying to achieve.

Some bending problems originate with the horse and are more embedded in the issues of balance, looseness, and straightness. Riding a horse who is stiff when tracking to the left, and some horses are naturally more stiff than others, can feel as though you're riding a 2×4.

The horse's entire body can feel as though it's falling in against your inside leg. Fixing this problem requires suppling exercises, plus a rider who can encourage a horse with inside aids to loosen and, over time, bend even though the muscles are resistant.

Another problem is the horse who tracks counter-bent, which is to say he's bent opposite to the direction he's traveling. This

horse is out of balance and falling in. The rider is trying to correct the problem, but is overusing the outside rein. The rider needs to reestablish diagonal aids, get the horse to move off the inside leg, and be balanced in the outside rein.

You can have flexion without bend, but you can't have bend without flexion.

▲ This horse is counter-bent, with his head bent left as he tracks to the right.

USING THE DIAGONAL AIDS

Flexion and bending form the core of diagonal aids. In the beginning of a rider's education and of a horse's training, the horse is being ridden using lateral aids: left hand for left work, right hand for right work.

With diagonal aids, riders and horses cross an important threshold. You are riding your horse from your inside leg into your outside rein. (For more on diagonal aids see page 46.)

#29 The Half Halt

SIX-TIME OLYMPIAN ROBERT DOVER once told a group of young dressage riders that he had two words of advice for every rider: half halt — golden words not just for the young dressage rider, but for every person settling themselves into a saddle.

Half halt is probably the most common command used by instructors in all corners of the world. You might also hear "adjustment" or "rebalance," but whatever the term used, this supremely important riding technique is often misunderstood. Of all the topics tackled in this book, the half halt offers, by far, the biggest challenge for the rider, in large part because it is so effective when used properly.

What Is It?

When I'm asked to define this term, I try to answer briefly and accurately so as not to totally mystify my student (enough of that will come later anyway). My description is that "a half halt is a little *whoa* (to your horse), followed by a little *go*, and finished with a little release." Easy enough? If only.

The FEI offers this on the subject: "Every movement or transition should be invisibly prepared by barely perceptible half halts. The half halt is an almost simultaneous, coordinated action of the seat, the legs, and the hands ... with the object of increasing the attention and balance of the horse before the execution of the movements or transitions to lower and higher paces.

"By shifting slightly more weight onto the horse's hindquarters, the engagement of the hind legs and the balance on the haunches are improved for the benefit of the lightness of the forehand and the horse's balance as a whole."

▲ FORWARD TROT

▲ MOMENT OF *WHOA*

An analogy that many students are able to understand is that a half halt is somewhat like a clutch on a car or truck. (With fewer and fewer people driving stick shifts, this analogy is becoming less meaningful, but I offer it anyway.) When you engage the clutch, you are telling the engine that you're either going to shift up a gear, down a gear, or maybe just stay in the same gear. The engine pauses when the clutch is engaged and waits for the driver to make a decision. This is the *whoa* portion of the half halt.

To further the analogy, the driver revs the engine when the clutch is engaged so that the car's energy is being prepped even before the next shift. This is the *go* part of the half halt. Then, when you shift the car back into a gear, the car continues. This is comparable to the release of the half halt.

Why Is It Important?

A half halt is such a make-or-break proposition that once you develop the skill and timing to use it well, you find you absolutely can't ride without it. The main purpose of the half halt is to control and direct the ever changing flow of energy between horse and rider. In directing and controlling the flow of energy, you can find a thousand and one reasons to use the half halt, and the more you use it, the better your ride will be.

When you use a half halt, you tell your horse that something might change. You may be asking him to go from one gait to another or to change tempo within a gait. You may want him to stay in the trot but be prepared to stay balanced through the upcoming corner. A half halt tells your horse to be ready for something new.

Here is a by no means complete list of the times within a ride that you could or should use a half halt:

▶ Transitions from one gait to another
▶ Shortening or adjusting strides within a gait
▶ Alerting the horse that a new exercise or movement is coming
▶ Improving or maintaining the horse's collection and self-carriage

The questions of frequency, intensity, and intention are associated with every half halt. The rider's use of it depends on the situation. Think of the half halt as having a volume control: some siginals are loud and quite obvious, others are soft and barely audible, and many are somewhere in between. Although Grand Prix riders in competition use beautiful, absolutely imperceptible half

▲ GO AND RELEASE

halts, during training there are plenty of times when a "loud" half halt is just what the doctor ordered.

What Happens in a Half Halt?

If a horse is moving along with a good forward energy in a trot or canter (walk is not necessarily the most productive place to half halt because there's minimum forward energy) and the rider asks the horse to slow down while at the same time asking for more energy and engagement, the horse should make several noticeable changes. He should:

▶ Feel lighter in the hands
▶ Show an increased desire to move forward
▶ Move more positively toward the bit
▶ Feel more longitudinally balanced

Aids for the Half Halt

Now for a look at appropriate aids. The process can be divided into three basic steps, but timing is *everything*.

THINK FORWARD. Start with the idea that your horse must be pushed forward. This action is done with a bit of seat and leg.

SLOW DOWN. At the same time, you want to send the message to briefly slow down. It's important that the hands not take over. Too much rein will block forward movement, and the horse will

stop moving the back legs (just the opposite of what you really want). You want to "enclose" your horse between your weight, leg, and rein aids to hold the horse for a brief moment.

SOFTEN UP. After your horse responds to your message of *wait*, the third and most important step is to soften your aids so that your horse is released from the hold and moves forward again.

Let me reiterate the words "frequency," "intensity," and "intention." The above description of the broad aids and their sequence opens the question of how often, how strong, how soft, and for what purpose?

A good suggestion when working on developing useful, correct

half halts with your horse is to play with the aids and their intensity. Find out what is effective and what isn't on *your* horse.

Every horse will respond differently to a combination of the weight, leg, and hand aids. The question is what combination you need for your horse.

Fixing Mistakes in the Half Halt

The biggest mistake seems to involve the hand. When a rider uses too much rein, the horse resists and either tosses his head or simply slows down. If a rider uses too much leg, the horse could speed up and the result would be to run onto his forehand.

A HALF-HALT EXERCISE

If you've not ridden half halts before or just need to tune your horse in to your aids, here is a great exercise that simplifies and clarifies.

Start your horse with a trot. Now, ask for a walk transition with the idea that your horse is listening to your seat and enclosing leg, and then the application of the hand aids. After just a stride or two of walk, ask for a trot again. Next, repeat the sequence a few times.

You'll quickly find that your horse starts to tune you in and is waiting for your next request. Next, ask for the walk, but at the *moment* your horse is going to walk, offer the release and some leg as if to say, "Oh, never mind about that walk. Let's just stay in trot." *That* moment is an example of a very loud half halt.

You can now try to become lighter and quieter with your aids to make the half halt less perceptible. You can also mix it up and add in a few full transitions to walk.

CHAPTER 7

SCHOOL FIGURES

Schooling figures form the core of training sessions for both horse and rider. Think of arena figures like deposits at a bank: the value you get out of the exercises is in direct proportion to the effort you put in. The precision of practice in the arena pays off in having a horse who is obedient, balanced, and a pleasure to ride. The essential goals of dressage are being achieved when the figures come together well.

The majority of riders have three basic riding environments to choose from: out in the open on natural terrain, in a maintained arena with consistent footing but of no specific size, or in a formal dressage arena with measured points and letters. The latter is the ideal setting for working on school figures, as it allows you to pay particular attention to the exact size and execution of each exercise.

This is not to say that you can't work on circles and other school figures in an unmarked riding space. It's just much more difficult to get the full value out of a 20-meter circle, for example, when you have no markers to clearly define the exact location of the points of the figure.

The Dressage Arena Explained

When you walk into a dressage arena, you might feel as though you're having a *déjà vu* moment of geometry class, with diameters, perimeters, circles, straight lines, diagonal lines, and so on. Once you feel comfortable in a dressage arena, however, the good news is that a correctly sized and marked arena is the same whether you're in the United States, Germany, Australia, Canada, or anywhere else in the world.

A standard-sized arena, also referred to as a "large" arena, always measures 20 by 60 meters. A schooling arena, or "small" arena, is 20 by 40 meters. (A standard-sized arena is assumed when discussing figures here, but a small arena still offers good references for precise schooling figures.) The

dressage letters that form the basis for your exercises are within this arena. The location of the letters is always the same and is measured in meters from exact points, such as the corners or centerline.

Purpose of the Arena

The purpose of the dressage arena and schooling figures is to help define the training goals and give valuable feedback to the rider. If you're beginning a circle at C, it should begin at that letter, not a meter before or after. I once heard a trainer say that jumping is all about the excitement and dressage is all about the precision. The arena makes the precision possible.

The quality of the figures you get with your horse tells you a lot about the quality of your riding. If your 20-meter circle ends up looking more like an egg, you know you're having trouble with aids, balance, obedience, and a host of other issues. It's almost impossible to get a precisely circular shape without having all the elements of a good ride in place.

Schooling Figures Checklist

In the early stages of learning to ride and train a horse, the arena figures are basic: circles, bending lines, and straight lines. These basic figures are the focus of this book, but as you advance, figures become more challenging as they change in size and begin to include nuances and variations that are limited only by the imagination.

As you work through the arena figures in your daily schooling, it's good to keep in the back of your mind a checklist of criteria or goals:

STEADY RHYTHM. Your horse should maintain a steady rhythm through all circles, turns, bends, and straight lines.

TRACK STRAIGHT. As shown on the right, your horse should be tracking straight, as if on a railroad track, in all circles, turns, bends, and straight lines. Remember the definition of straightness from page 107: "Straightness is when a horse tracks with his forehand in line with the hindquarters on both straight and bending lines and in all three gaits."

WORK BOTH SIDES. Make sure to work your horse in both directions equally to build balance between the two sides of his body.

Working on Circles

The circle is the most useful schooling figure of all. It is central to working with young horses and new riders both on the lunge and under saddle. You can ride a circle in the walk, trot, and canter. Circles are not something you grow out of; the most highly trained horses and the most experienced professionals use them every day. You see circles in dressage competitions, in training, and in jumping. Where would riding be without the circle?

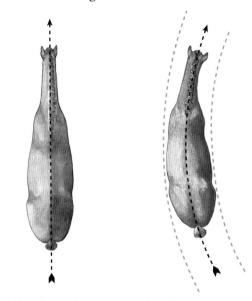

▲ Tracking straight, even on a bend

ARENA LETTERS — THE ALPHABET GONE MAD

Riders new to dressage always wonder what the letters around an arena mean and try to find some logic. The bottom line, I'm sorry to say, is that the letters mean nothing and there is no logic. Letters were not even used in international competitions until the twentieth century. The first Olympic competition in 1912 did not have letters in the arena. For some unknown reason, the letters showed up in the 1920 Olympics.

The FEI offers two possible explanations. One theory holds that the letters are the first letters of the names of cities conquered by the Romans. The second is that, historically, the court walls in a German royal stable were marked with letters for locations where the members of the court were to collect their horses. For example, K is for the King (or Kaiser) and H for the *Hofmarschall* (or Lord Chamberlain). This theory has some holes in it, since in the 1920 Olympics, the centerline was also marked with letters, and in royal times there was no designation for the middle of the court.

Whatever the history, some people use the mnemonic "A Fat Black Mare Can Hardly Ever Kick" to mark the letters at the corners, ends, and midpoints. The letters that fall between the endpoints and the middle are easy to remember too: RSVP. The centerline letters, however, have no rhyme or reason: GIXLD.

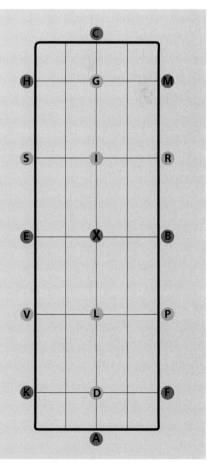

#30 The 20-Meter Circle

YOUNG HORSES, older horses being rehabilitated, and horses being schooled for a new discipline or coming back into training after a hiatus should always be started on a 20-meter circle, nothing smaller. The newer rider also needs to master the 20-meter before going any smaller. As the horse develops more muscle and better balance and the rider gains more skill, a smaller, more demanding circle can be ridden.

A 20-meter circle is traditionally ridden in three different places in the arena: a circle starting at A; a circle starting at C; or a circle in the middle, with letters B and E marking either the beginning or the end. The arena walls and markings make it easier for horse and rider to be accurate in these locations.

If you ride a circle at either end of the arena (A or C), you have the advantage of three walls to guide you. The circle in the middle offers only two walls and can sometimes be a little trickier. There is no error, however, in doing a 20-meter circle at any point within the arena in your daily training. As long as you start and end your circle in the same spot, hit the correct opposite spot on the far wall, and create a round figure, it's good training.

At the beginning levels of dressage competition, the entire test is built around 20-meter circles because it's so much easier to keep a horse focused, supple, and on the rider's aids when he is on a circle. As soon as you go off the circle and head straight, a horse becomes harder to ride, by which I mean that the circle itself has the virtue of helping to keep the horse supple, bending, and connected. On the straight parts, a rider has to use more skill and correct aids to achieve the same results.

20-METER CIRCLE ▶

How to Ride the 20-Meter Circle

While a 20-meter circle is the easiest circle to ride, mastering it is another matter altogether. Why are circles so darned hard? It comes down to the fact that every footfall presents an opportunity for an issue that needs correcting. In riding a circle, the first thing to understand is that the arc of the horse's body needs to match the arc of the circle.

To ride a circle correctly, a rider needs to have correct flexion and bend with fully functioning diagonal aids, as described here:

START BY SLIGHTLY WEIGHTING the inside seat bone.

USE YOUR INSIDE LEG to ask your horse to bend.

USE THE OUTSIDE LEG as a guarding aid to keep the circle from losing its shape to the outside.

USE YOUR INSIDE REIN to ask for flexion to the inside.

USE YOUR OUTSIDE REIN to support the horse.

When teaching someone how to ride a circle, I like to ask what sounds like a trick question: "How many strides of 'straight' are on a circle?" The correct answer is none, of course, but it is the rare circle indeed that isn't ridden without at least one stride of straight somewhere when you are first learning.

Riding a *round* circle is a matter of being a proactive rider as opposed to a reactive rider. A proactive rider plans ahead and foresees what problems are likely to be encountered. If you have not tried to ride a geometrically correct circle, or if you always struggle, setting up cones to help you train is always a great way to get it right.

For this example, we'll ride a 20-meter at A, but the other traditional starting points work well, too. Begin your circle at A, and through the entire circle keep your eyes looking ahead to the next cone. As you approach a cone, make sure you look ahead to the next one and even beyond. This training exercise gives both you and your horse the idea of correct size and shape for the circle, and how you need to plan ahead to get that shape.

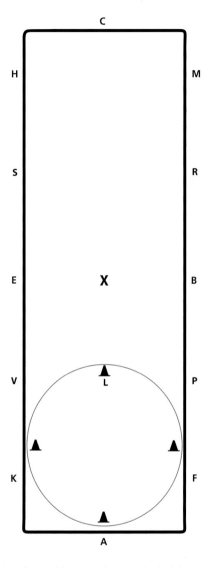

▲ Start with a cone just on the inside track at A. Letter L is directly opposite A, on the centerline. Because you want your horse to track two meters past letter L, put your cone at *one* meter beyond L, and make your circle on the outside of that cone. The next two cones go inside the track, four meters down the long wall from letter K and four meters down the wall from letter F.

Fixing Problems with the Circle

There are many potential pitfalls in riding a circle. A common error is when the rider drifts into the corners of the arena, so that the shape of the circle stretches. Although riding corners is an equally important element of correct riding, you don't ride corners on a circle. Use your outside leg and rein to keep the horse on the curve of the circle.

Focus on Turning

Another routine error occurs when a rider forgets to tell the horse that he is turning. The best example of this is when a rider begins a circle at A. As long as she is able to use the wall as a guide, things go pretty well.

The circle falls apart, however, when she asks the horse to continue to turn and go into the open space of the arena. There are often about five strides of straight before the rider sends the message to the horse that he needs to continue on the turn of the circle.

Every stride on a circle is a turning stride.

There's a glue that seems to bind the horse to the wall, and leaving it is like a fly trying to get out of sticky tape. The good news is that overcoming that sticky tape is great training for both horse and rider. Riding ahead with your eyes and keeping in mind that every stride is a turning stride will do wonders.

Stay Balanced

Another common issue is the quarter of the circle just past the centerline, as the rider approaches the oncoming wall. This is where the horse often becomes unbalanced. The outside shoulder pops out of alignment and starts pulling in the other direction. To address this, you need a consistent outside leg and rein.

The opposite issue appears when the horse loses balance and collapses in on the circle. The overuse of the inside rein can cause either of these problems. A rider needs to use her inside aids to keep the horse balanced into the outside rein. (See Diagonal Aids on page 46.)

There are many terms used to describe these issues: falling in, falling on the inside shoulder, falling out, falling over the outside shoulder, bulging in, bulging out, leaning in, leaning out, collapsing in. All indicate that the horse has lost balance and is no longer tracking straight. Remember from the training scale: your horse should always be tracking straight (front legs in line with the back legs as though on a railroad track), whether on straight, bending, or circle lines.

CIRCLE INSIGHT

In a rider's education there is a "light-bulb moment" when you first ride a circle with the correct use of aids. It makes intuitive sense that you would "steer" a horse through a circle by using your inside rein, but it's wrong. You certainly would with a bicycle, but it's wrong for a horse.

For a horse on a circle, the correct aids are counterintuitive: the inside aids give a horse flexion and bend, and the *outside* aids turn him. Badly shaped circles are quite often a consequence of a rider turning the horse using the inside rein as though the horse were a bike, instead of using the outside rein and outside leg as guiding and supporting aids and the inside leg as an encouragement for bending.

#31 The 15-Meter Circle

ONCE THE 20-METER CIRCLE becomes easy for both horse and rider, the pieces of the puzzle should be in place for a 15-meter circle to be ridden with ease.

A good reference point for figuring a correctly sized 15-meter circle is to do a circle at E or B, and make the top of your circle point the quarter line, since that's 5 meters in from the outside wall. The entire circle needs to shrink down from a 20-meter accordingly, so that you're not making an oval or egg-shaped figure. As always, the arc of the horse's body needs to match the arc of the circle, but now the arc is more accentuated.

A great rule to keep in mind: your horse should always be looking in the direction he's going.

15-METER CIRCLE ▶

▲ Although the issues are the same as those with a 20-meter circle, they increase in intensity because the circle is smaller, the bend is greater, and so on.

#32 The 10-Meter Circle

THE 15-METER CIRCLE is really just a stepping-stone to the next big gymnastic exercise in arena figures, which is the 10-meter circle. Ten-meter circles are not good for a horse just beginning his training. Because they're demanding of a horse's balance and athletic ability, he needs to be built up before riding a smaller circle.

▼ The 10-meter circle is a valuable training tool for working on collection and bend. The greater the degree of bending, the greater the gymnastic effect.

10-METER CIRCLE ▶

The 10-meter circle covers only half the width of the arena. If you're doing a 10-meter at E or B, the center letter X marks the top of your circle. The degree of bending is quite dramatic, and the challenge of balance is no small matter. Just as a 15-meter circle is harder than a 20-meter, a 10-meter is even harder for both horse and rider.

Ten-meter circles can be used anywhere in the arena and are useful for getting a horse to engage the hindquarters, working on collection, and moving into more advanced exercises like shoulder-in and haunches-in. (In the first stride of a 10-meter circle, the horse is in a position that represents shoulder-in, and in the last stride, he's in a position that represents haunches-in, but that's for another book!)

Circles are not something you grow out of; even the most highly trained horses and the most experienced professionals use them every day.

WHAT'S A VOLTE?

A volte is another of the many riding words that seems to create confusion and disagreement. Some trainers will say it's a 6-meter circle. Others will tell you it's a 6-, 8-, or even a 10-meter circle. Still others will tell you it's anything smaller than a 10-meter – including 6-meter and 8-meter circles. What is the right answer? The FEI states clearly that the volte is a circle of 6, 8, or 10 meters. If it's larger than 10 meters, it's called a circle. So here we are considering volte to be a catchall term for a small circle.

As you move up the levels with your horse, you are asked to trot and canter an 8-meter circle, which is the smallest volte asked for in competition. A 6-meter circle is the smallest circle you can ask a horse to do, because biomechanically the horse cannot bend enough through his body to stay on the track of a smaller circle without losing the hindquarters.

#33 The Half Circle

A S ARENA FIGURES continue, the natural progression is from circles to half circles. Gymnastically, you want to make sure you're working your horse evenly on both sides, and a half circle presents a great way to have your horse change directions. Other common descriptive terms include "half turns"

and "teardrop turns," because the figure ends looking like a big teardrop.

As with all dressage training, the half circle is a building block for more advanced development of the horse. For example, the half circle can be modified so that a rider leg yields back to the track. As the horse progresses, this

exercise can also be used to teach a horse how to half pass.

The half circle shouldn't be viewed purely in the context of dressage. Jump courses are full of them, with more advanced courses calling for smaller and smaller ones. A rollback, in jumping, is nothing more than a half circle between one jump and the next.

▼ 10-METER HALF CIRCLE

How to Ride a Half Circle

As with circles, half circles traditionally happen at a couple of specific places in the arena.

Option 1:

▸ Take your horse down the track to E if you're tracking right or B if you're tracking left.

▸ When you arrive at the middle letter, make a half circle of 10 meters to letter X.

▸ Instead of continuing your circle at X, head back to the corner on the same side of the arena that you just left, in this case K. If you're tracking left, the line is from X to F.

Option 2:

▸ Ride a half circle in a corner at the end of a long side and arrive back on the track on the opposite rein.

There is nothing wrong with doing the half circle anywhere along the long side of the arena as long as you leave yourself enough room to return to the track in the opposite direction.

Crookedness on the Half Circle

As a consequence of a horse's natural crookedness, a half circle in one direction will always feel a little harder than in the other. (See the straightness discussion on page 107.) For the majority of horses, the half circle to the right can be hard to keep to 10 meters, as the horse will shift and lose balance to the left.

When the horse is turning to the left, his natural stiffness in that direction can make getting correct flexion and bend a problem. Suppling the horse for more equal bend and flexion is precisely why the half circle exercise is so good for him.

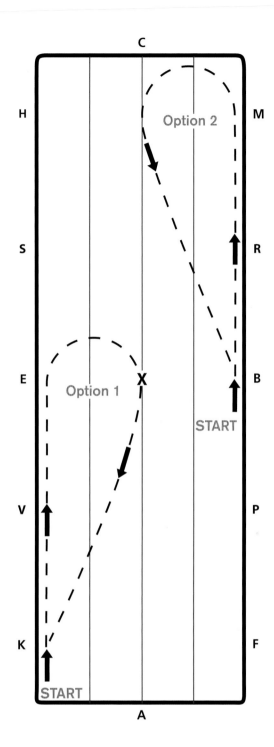

#34 Figure Eights

I F ALL THESE ARENA FIGURES are starting to remind you of figure skating, then figure eights will really make you think you should put on your skates. This exercise is similar in nature to the half circle in that it provides a change of direction and, therefore, gymnastically works both sides of the horse.

The figure eight is used in basic training and as part of competition. The main challenge with a figure eight is the correct change of bend and flexion in the middle of the figure.

How to Ride a Figure Eight

The figure eight is an extremely versatile exercise that can be ridden many different ways.

Perhaps the most frequently used version (1) is the figure eight created by two 20-meter circles with letter X as the center. You must master a single 20-meter circle before trying to glue together two in opposite directions. Unlike the other arena figures described so far, where no part of the figure is straight, this

type of figure eight does have a stride of straight that falls right at the point where the two circles meet.

To ride it correctly, give the aids for a circle in one direction, half halt to prepare for a coming change, straighten the horse for one stride, and then ask for bend and flexion in the new direction. The straight stride (which in the beginning may take more than one stride) is imperative. Your horse needs an opportunity to rebalance and prepare for a new direction before going there.

▼ BASIC FIGURE EIGHT

FIGURE EIGHTS AT THE CANTER

Cantering a figure eight is more complicated because you typically have to deal with a change of lead between the two circles. (A more advanced horse might be in true canter for one circle and counter canter for the other). A change of lead through the trot is the easiest choice when riding a cantering figure eight. This is where the horse is brought back into the trot for a few strides and then picks up the canter again with the new lead.

Progressive training would then move to a simple change in which the horse is brought back to walk directly from the canter. Then, after a few clearly defined steps of walk, the horse is asked to canter again on the new lead, with no steps of trot. Eventually, a horse can do a "flying" change of lead, which means no strides of trot or walk when changing direction.

Figure eights do not have to be 20 meters in size. You can ride 15-meter figure eights and then progress to 10-meter figure eights. The 10-meter version can be ridden using letter X facing B or E (2) or using the centerline (X or any point along the centerline) as the one stride of straight (3).

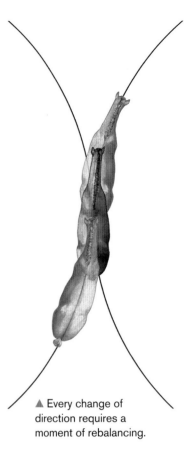

▲ Every change of direction requires a moment of rebalancing.

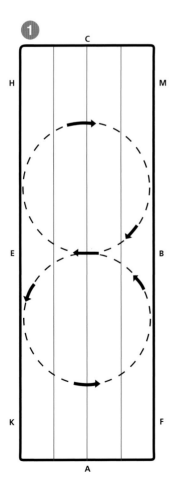

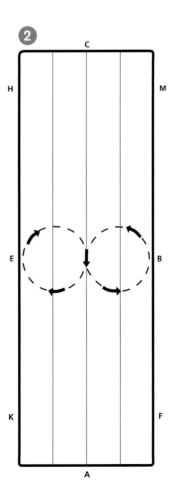

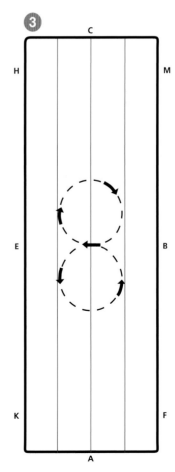

#35 Bending Lines

WHEN I FIRST introduce bending lines and serpentines, I like to ask my riders to guess what the root word of the term serpentine is. As they usually know, a serpentine is a schooling figure that looks like a snake through the arena. There are many types, sizes, and variations.

The FEI describes it this way: "The serpentine with several loops touching the long side of the arena consists of half circles connected by a straight line. When crossing the centerline, the horse should be parallel to the short side. Depending on the size of the half circles, the straight connection varies in length. Serpentines with one loop on the long side of the arena are executed with 5 meter or 10-meter distance from the track. Serpentines around the center line are executed between the quarter lines."

Schooling figures are building blocks, and serpentines are one of the best. They can be ridden in walk, trot, and canter, and in working and collected gaits. They are brilliant for building balance and suppleness in your horse. And they are easy to master because they build on what you've already accomplished in riding correct circles, half circles, and figure eights.

Remember that these arena figures are not just something you do in dressage. If jumping is your goal, you'll easily see how establishing these curving lines on the flat can translate into well-ridden jump courses. Consider the event horse as well. Many event courses ask for horse and rider to snake their way through a series of obstacles, and serpentines work directly on these skills, too.

Starting with a Single Loop

The beginning of serpentine work often is a single-loop serpentine, also commonly referred to as "riding a single-loop bending line." (Multiple names for the same thing seem to be the rule rather than the exception in the much cross-referenced world of horses.)

Some trainers refer to this version of serpentines as bending lines or broken lines. Whatever the label, this version of the serpentine is the second part of the FEI definition: a serpentine that makes a single loop away from the wall.

The single-loop exercise, which asks the horse to leave the track and return to it without changing direction, is the easiest for a young horse to manage or a new rider to work through. The gentle bending off the track and back does not pose a high level of difficulty in regard to balance and steering.

The goal is to help a horse develop better balance and become more supple as a consequence of tiny changes of bend and flexion. A horse's straightness also benefits because he's asked to keep a straight body even though he's on a bending track.

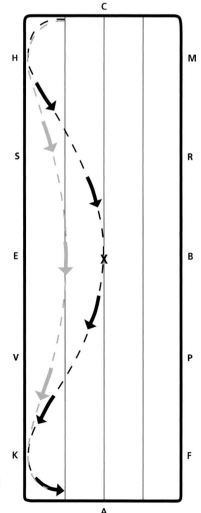

▶ SHALLOW BENDING LINE (GREEN)
DEEPER BENDING LINE (BLACK)

How to Ride a Single Loop

To begin a single-loop serpentine, you should ride your corners (see page 148) to prepare your horse to bend softly off the wall at the corner letter. If you use your corner constructively, setting your horse up to leave at the corner letter is just a gentle turn.

There are six meters between the corner of the arena and the corner letters — M, H, F, or K. Use that six-meter space along the wall to prepare your horse to leave at the corner letter.

Missing the corner leads to something akin to a hairpin turn when you try to leave the track at the corner letter. By missing the corner itself, you've used up most of the six-meter space you could have used to prepare for the movement away from the wall.

From the corner letter, your horse should be flexed and bent in the direction that you are going. Ask him for a shallow curve that peaks at the quarter line (5 meters off the long wall). When you reach the quarter line, you need to change the bend and flexion, and ride toward the next corner letter on the side you just came from.

The bend in your horse's body for this shallow bending line is minute, but very important. (It will become more pronounced as the bending line becomes more dramatic when you progress to steeper bends in your serpentines.)

As you hit your new corner letter, the bend and flexion change once again, and you need to ride from the corner letter to the corner itself to correctly finish the bending line.

Deeper Bending Lines

Once you're comfortable with the shallow bending line version of the serpentine, the next step is to take your horse deeper into the arena for a one-loop bending line to the centerline.

The preparation is the same as for the more shallow lines. The movement itself asks a horse to go to the point of X before changing the bend and flexion and then return to the track by the next corner letter. This particular bending line can play mind games with a rider's balance as well because you're typically prepared to change directions by the time you get to X. The balance problem posed here is that you do not change directions even though you push it right to the edge.

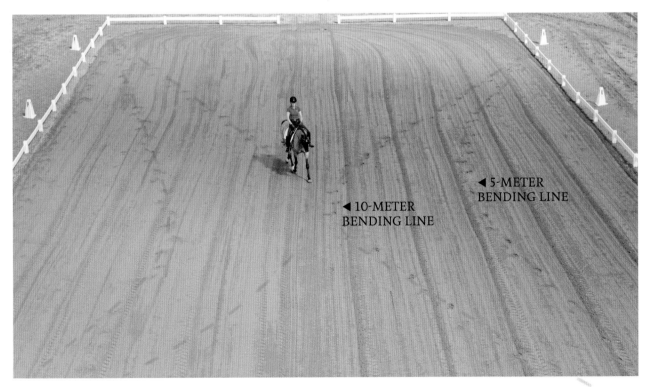

◄ 5-METER BENDING LINE

◄ 10-METER BENDING LINE

▲ The single-loop bending line marks should be 5 meters from the track for a shallow line and 10 meters for a deeper one.

Double-Loop Bending Lines

Another variation on the serpentine theme is a double-loop shallow bending line (right). As in the single loop, you set your horse up for a soft curving line to leave the track at the corner letter, but when you reach 2.5 meters out from the wall, you softly bend your horse back to the track,

arriving by the middle letter (E or B). At that point you ask for flexion and bend to the inside again, then leave the track by about 2.5 meters, and return by the corner letter.

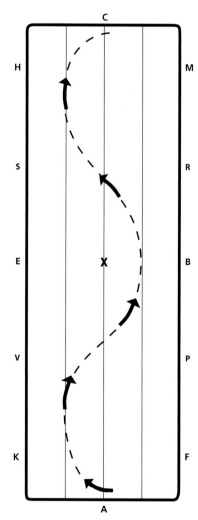

▲ In this version, you ride the serpentine line from quarter line to quarter line. The immediate value of this particular form is that there are no walls for your horse to balance against; your aids have to form the balance for the horse. For the horse who falls to the wall, this is a helpful exercise for redefining and clarifying aids and balance.

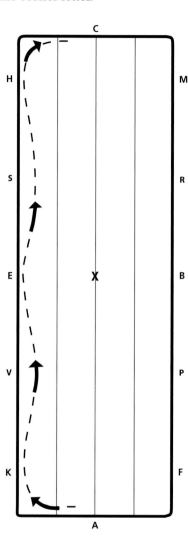

Common Problems with Bending Lines

The mistakes commonly seen with these early serpentines begin with not riding the corners properly. In riding all arena figures, it's important to keep in mind that your horse needs to be balanced using diagonal aids. Riding the beginning of your bending line as though you're on a circle does not set your horse up for success, because the true challenge is keeping him straight while both leaving *and* returning to the track.

Another issue you'll often see is a horse returning to the track as though being asked for a lateral movement. This unasked-for "leg yield" is usually accompanied by a popped outside shoulder and a rider who is trying to fix it by overusing the inside rein, which results in an overflexed horse who falls to the wall.

One fix for both problems is to slow down and do the movement at the walk, giving you — and your horse — the chance to see where the problems are in balance and straightness.

DEALING WITH THE DOMINO EFFECT

In riding, as in so many endeavors, one mistake leads to another and another, and it's not long before the whole intent of an exercise has dissolved into a mess. If you find yourself in this place while riding arena figures, often the best solution is to slow down and walk through the figure. This process gives you a chance to think through the aids and where you need to help your horse along. After you get a feel for the correct movement at a walk, go back to the trot and try again.

#36 Three-Loop Serpentines

THE BENDING LINE is the first, easier bending task for horse and rider. The three-loop serpentine makes that task more complex and difficult because of the number of turns and changes of direction. It's a fantastic suppling exercise, working your horse in both directions, back to back. It forces you to focus carefully on balance, again as the line shifts from left to right and back again.

How to Ride a Three-Loop Serpentine

The most common method of riding a three-loop serpentine is to not ride the corners and have only one stride of straight between the loops. This version starts at one end of the arena (A or C), and begins as though you are going to do a 20-meter circle. As in the figure eight, when you hit the centerline, you ask your horse to be straight for one stride, change the flexion and bend, and start a 20-meter circle in the other direction.

When you complete just half of that next 20-meter circle, again at the centerline, you ask for one stride of straight, change the flexion and bend, and then make another half of a 20-meter circle, again in the opposite direction. The figure ends at the side of the arena opposite from where you started. If you were to continue the exercise, you could complete

▼ THREE-LOOP SERPENTINE

all of your 20-meter half circles by riding the serpentine back up the arena to where you started.

This wall-to-wall serpentine can also be ridden to include four half circles of equal size, making it a more advanced exercise. The serpentine will finish on the same rein if there's an odd number of loops, and will finish on the opposite rein if there's an even number of loops.

Ribbon-Candy Serpentines

A variation on the theme is an old-fashioned serpentine or, perhaps more descriptively, the "ribbon-candy serpentine," where each loop swings into the next without any strides of straight. The centerline is ridden at an angle and the loops curve back on themselves, making tracks reminiscent of ribbon candy. The horse's flexion and bend change smoothly at the point of the centerline. This exercise is a good way to make your horse more supple and loose.

As your horse advances, ride your serpentines in canter, at first with alternating loops of true canter and counter canter, to simple changes (canter, walk, canter), and, finally, with flying lead changes.

Fixing Serpentine Problems

One of the biggest problems with riding serpentines is equally dividing your loops and making them symmetrical. You need to mentally divide the arena up into its geometrical parts so that each loop is the same size. For those who have spatial-relations difficulties, this can be a true challenge. As when figuring out how to ride correct 20-meter circles, cones can be a huge help. Place cones in the center of the arena to mark where you should make your loop and guide you through the exercises.

Another common mistake is when your horse is not tracking straight (horse's front feet in alignment with his hind feet) on what turns out to be a constantly bending line. If this problem crops up, consider going back to circles to focus on getting the horse on the circle's line in one direction at a time.

▼ RIBBON-CANDY SERPENTINE

#37 Long and Short Diagonals

THE LAST CATEGORY of schooling figures is diagonals, also called diagonal lines. If you've been reading this book sequentially, you've read about posting diagonals and diagonal aids; arena diagonals are yet another way that "diagonals" show up in riding.

How mystifying it would be for a nonhorse person to walk into an arena and hear an instructor say — quite correctly — "Let's change directions on the diagonal, don't forget to change your posting diagonal, and confirm your horse is in balance in the corners with your diagonal aids."

As the name suggests, diagonal lines cut diagonally across the arena. Using diagonal lines is just one of the many ways to change direction in the arena. We've already described half circles, figure eights, and serpentine loops. Now you can add long and short diagonals to the list.

Why Diagonals Are Important

Although diagonals may sound simple on paper, they can present surprising challenges to the horse and rider. Dressage is all about precision. Riding extremely accurate and correct diagonals is not only vitally important if you expect to do well in a dressage competition, but essential to teaching a horse about straightness and balance for any discipline. If you do any jumping, for example, you will commonly find a sequence of jumps set up on a diagonal in the show ring.

The diagonal line forms an important educational and training step in dressage. Although the lower level tests ask simply for

▼ LONG DIAGONALS

accuracy at the letters and straight lines between the letters, at higher levels the diagonal becomes the proving ground for trots that show lengthenings, mediums, and extensions, and canters that demonstrate mediums and extensions, as well as flying lead changes, pirouettes, and tempi changes. Riding the correct, precise diagonal provides the foundation for all these other movements.

Riding Long Diagonals

For long diagonals, use a "far" corner (meaning the second one on the short side) as your starting point. Leave the track at the corner letter, cross the arena in a straight line that goes directly over letter X, arrive at the opposite corner letter, and then ride the corner. There are four correct ways to ride a long diagonal: H to X to F, F to X to H, M to X to K, and K to X to M. If you rode one long diagonal line after another, you would be completing a full-arena version of a figure eight.

Riding Short Diagonals

The short diagonals offer more variety. A short diagonal uses just half of the arena to change directions. There are four ways to cross the arena if you're using the corner letters to the middle letters. The options are F to E, M to E, K to B, or H to B.

There are also four possibilities using the middle letters as starting points and ending up at the corner letters. These are B to K, B to H, E to M, and E to F.

What's the Difference?

What's the difference, you may be wondering? There are times when it makes good sense to ride a short diagonal versus a long one. For example, you can use a short diagonal if your horse gets too quick on straight lines, loses longitudinal balance (falls on forehand), or otherwise becomes distracted. Changing directions across the wide-open space of the long diagonal is sometimes more than you want to ask of a young horse or a newer rider.

Finally, it can be too much to ask an inexperienced horse who is learning to lengthen the stride in trot to consistently maintain the propulsive power on the long diagonal. The short diagonal gives the horse a chance to maintain power to the end of the line without being overtaxed.

▼ SHORT DIAGONALS

How to Ride the Diagonal

Riding a correct diagonal for any of the long diagonals and for half of the short diagonal versions starts with a good corner (see pages 148–149), as described below:

WITH THE HORSE BALANCED through the corner, position him to the inside as you are just approaching the corner letter. (Don't forget that you have 6 meters between the actual corner and the corner letter; use that room to position your horse.)

LOOKING ACROSS THE DIAGONAL line to the far corner, put your outside leg on the horse and ask him to leave the track.

GUIDING YOUR HORSE between your legs and hands, make an absolutely straight line right through the arena to the opposite corner letter (if you're on the long diagonal), cutting the arena into two halves.

AS YOU CROSS THE CENTER-LINE (precisely at letter X), switch your aids for a new inside leg, inside rein, outside leg, and outside rein. This is subtle, but important if you're going to have your horse balanced in the new direction. If you're in a rising trot, change your diagonal. (Some trainers look for the change of posting diagonal at the new corner letter.)

WHEN YOU ARRIVE at the new corner letter, if you're riding correctly, your horse's nose is the first to arrive at the wall, not the side of his face or his shoulder. Ask for the new change of bend and flexion and ride the next corner using your diagonal aids.

Fixing Problems with Diagonals

Most problems associated with the diagonal come down to a lack of precision. It takes advance planning, attention to detail, and good use of a rider's aids to leave the track at the corner letter, cross directly over letter X, and arrive at the track at the new corner letter.

If you use too much inside rein when asking your horse to come off the track at the corner letter, it's likely that the horse will pop the opposite shoulder and be out of balance. You'll end up with a crooked, rather than straight, line. This quite often leads to an overcorrection, and then the diagonal line looks like a failed sobriety test. I call it "drunken sailor syndrome."

Using your eyes to focus yourself on the line ahead will help keep you precise. For example, if you're crossing the HXF diagonal, as you turn at H, look toward F. Any time you feel your horse drift from that line, make a tiny correction with your aids.

▲ It takes concentration and small corrections with the aids to keep your horse moving straight through the entire length of a long diagonal.

#38 Riding Corners

"RIDE! YOUR! CORNERS!" Any person who's spent any amount of time learning the precision of dressage has heard those words. Regardless of your level of riding, you need to always be focused on riding your horse through the corners of the arena.

The difference between training level and Grand Prix is just how deep into the corner the horse is ridden; the more advanced the horse, the deeper the corner. The horse is always, however, balanced with an even rhythm in whatever gait you happen to be in.

Almost everyone begins their riding experience by riding the short side of the arena as if it were a half circle, essentially ignoring the 90-degree angle of the arena ends. The corners, however, offer excellent training opportunities. By asking your horse to step into the corner, correctly flexed and correctly bent, you give him a natural opportunity to step more under himself with his inside hind leg.

Why is this important? Because it works the connection between the hind legs and the reins, it's a chance to develop engagement. (For more about engagement, see page 114.)

◀ This rider has set up her horse to ride this corner well, but could be slightly deeper into the corner at this point.

Getting the most out of your corners requires a quick sequence of aids as you go from long side to short side to long side of the arena. When you're actually riding it, the aids come too fast for an instructor to dictate; you'll need to know them on your own. The first stage is getting from the long side to the short side:

START WITH A HALF HALT at the corner letter to prepare your horse for the corner. (Remember that the corner letter is actually 6 meters out on the long side.)

CONFIRM THAT YOU'VE GOT inside flexion and simultaneously push with your inside leg to ask your horse to step underneath himself with his inside leg. This will push your mount into your outside rein.

THEN SOFTEN WITH the inside rein, allowing more inside bend through the horse's body. At this point, you'll be coming out of the first corner, starting along the short side of the arena. You'll need at least one and possibly two half halts along the short side before you start into the second corner, where you use the same sequence of aids as in the first.

WHEN YOU COME OUT of the short side, you need one more half halt to set up your horse for the movement down the long side, whether it's just riding straight or entering a movement such as a change of direction on a diagonal or riding a bending line.

▼ This rider is letting her horse fall into the center away from the corner.

CHAPTER 8

LATERAL WORK

While there are a multitude of terms that describe some sort of lateral motion, the most universal definition is having your horse simultaneously move forward and sideways.

WHEN LEARNING HOW TO RIDE a horse correctly and in training a young horse, the early focus is on forward, longitudinal motion. Lateral movement is the next natural stage of both riding and training. Because this is a book on fundamentals, we discuss lateral work at the most basic level.

Lateral work is important on many fronts. First, it's a great way to build your horse into a better athlete. Consider it yoga for your horse. Lateral movements also train the horse to be obedient and responsive to the aids, particularly the seat and leg aids.

Lateral Terminology

Within the definition of lateral work are five exercises: leg yield, shoulder-in (with shoulder-fore and "in position" as preparation for shoulder-in), travers, renvers, and half pass. Leg yield, however, is somewhat like the black sheep of the family and must be considered on its own terms. See Fundamental #40 (page 154) for more on that.

For now, let's begin with a key to all the terms that you're likely to come across in the broadest discussions of lateral motion, and how each is classified.

Lateral Exercises

▶ **Turn on the forehand** — A very basic movement in which the horse steps his hind legs around his inside front foot without moving forward. It is Fundamental #39 (page 152).

▶ **Turn on the haunches** — An intermediate movement in which the horse steps his front legs around the hind legs while the inside hind leg steps up and down, and the outside hind leg makes a small half circle.

▶ **Leg yield** — The horse flexes slightly away from the direction of travel and moves forward and sideways simultaneously, with one lateral pair of legs crossing the other. It is Fundamental #40 (page 154).

▶ **Shoulder-in** — The hind feet stay on the track while the forehand is moved from the track to the inside, making the horse's outside shoulder in line with his inside hip. The horse is moving on three tracks at an angle of 30 degrees to the track.

▶ **Shoulder-fore** — A subcategory of shoulder-in, and a preparatory exercise for all the lateral movements

▶ **Position in/In position/First position** — A sub-category of shoulder-in and shoulder-fore

▶ **Travers (haunches in)** — Intermediate/Advanced

▶ **Renvers (haunches out)** — Intermediate/Advanced

▶ **Half pass** — Advanced

Other Lateral Exercises

▶ **Side pass** — The horse moves sideways but not forward. Not seen or used in dressage.

▶ **Full pass** — In some older publications on classical riding, full pass seems to mean a half pass that uses the entire arena diagonal.

#39 Turn on the Forehand

ALTHOUGH NOT technically a lateral movement, a turn on the forehand is often one of the first lateral motions we ask of a horse. There is no call for a turn on the forehand in competition, and its place in lateral exercises is purely for gymnastic value and clarification of aids.

What Is It?

In a turn on the forehand, the horse steps its hind legs around its front legs, which are stepping up and down on essentially the same spot. The horse is flexed in the direction of the rider's asking leg, but he is not bent. In the early stages of training this exercise, it's common to ask for it while facing a wall so that the horse can't go forward.

With this method, the turn on the forehand is asked in a 180-degree configuration so that the horse is changing direction (along the wall) by having the hind legs step around the front legs. A turn on the forehand against the wall is the same movement for the horse as doing the turn in the middle of the arena, but it's easier for the rider because the horse cannot move forward.

As an introductory exercise, a turn on the forehand helps your horse understand that, although the weight aids and leg aids often mean "go forward," the same aids — differently applied — can also mean "go sideways." A turn on the haunches, which you might think should be in the same category as turn on the forehand, is considerably more complicated.

Why Is It Important?

The goal of a turn on the forehand is to help your horse learn how to move away from (yield from) a sideways driving leg aid. The exercise is a good tool for developing looseness in a horse as well. The movement is also highly valuable for teaching a rider how to coordinate the aids for lateral motion.

How to Do a Turn on the Forehand

A turn on the forehand can be done from a walk or from a good halt. In either case, follow this step-by-step guide:

FLEX. Ask your horse to flex in the direction of your sideways-pushing leg by using your rein aids, asking and yielding with that hand.

STEP. Ask him to step with his inside hind foot in front of and across the other hind foot. You should be asking with the leg on the side of the horse that's showing flexion.

PUSH. In pushing your horse's haunches to the side, use restraining rein aids to keep him from moving forward.

STEP. The front legs should be turning in small steps, as opposed to staying planted and pivoting on the ground.

Fixing Mistakes in Turn on the Forehand

The most common mistake when trying a turn on the forehand is that the horse steps forward or backward in an effort to avoid the turn. Given a choice between these two mistakes, stepping forward is always better than having a horse step backward.

Rushing the movement is another issue. It's not a race against time, so make sure each step your horse takes is a measured step of sideways motion with the haunches while the forehand actually steps up and down in place.

Occasionally, the rider offers conflicting aids — a leg that the horse interprets as a driving leg in conflict with restraining rein aids. If the rider is offering conflicting aids, it's often a sign that she is rushing through the movement faster than she can manage it, even if it's not rushing her horse.

In all of these problems, the rider's hand is often to blame: it's very easy to overflex the horse and try to pull him around with the inside rein.

FAR LEFT: The rider flexes her horse to the right and uses her right leg to ask the horse to step to the left.

CENTER: The rider controls the horse's desire to move forward with soft restraining hands.

RIGHT: The rider encourages the horse to step around to the left with the hind legs while the front legs step in place.

#40 Leg Yield

ONCE A HORSE STARTS to realize that moving sideways is an option, the natural training progression is to move into leg yield. Trainers the world over use the leg yield as the beginning of lateral work. The leg yield is the most basic of the lateral movements; the other four (shoulder-in, travers, renvers, and half pass) are more intermediate or advanced work.

The FEI defines the leg yield as follows: "The horse is almost straight, except for a slight flexion at the poll away from the direction in which it moves, so that the athlete (rider) is just able to see the eyebrow and nostril on the inside. The inside legs pass and come in front of the outside legs. Leg yielding should be included in the training of the horse before it is ready for collected work. Later on, together with the more advanced shoulder-in movement, it is the best means of making a horse supple, loose, and unconstrained for the benefit of the freedom, elasticity, and regularity of its paces and the harmony, lightness, and ease of its movements."

Why Is It Important?

The leg yield is a foundation principle that can be used in countless ways on the flat, in the fields, and when jumping. As you can see from the FEI definition, leg yielding can help your horse develop looseness. It is also wonderful for teaching a rider how to coordinate the aids and as a check for being truly between the aids with regard to straightness.

In the orderly process of training a horse, it's important that he be well versed in leg yielding

▲ Notice how the inside hind leg is stepping under and past the outside leg.

before being asked for more advanced lateral movements, all of which require a degree of collection. Leg yields also ask for activation of the horse's inside hind leg, which encourages him to step more underneath himself toward his center of gravity.

Variety in Leg Yield

Leg yielding can be done in many different ways. The list includes leg yielding from a line in the middle of the arena (quarter line, centerline) to the wall or from a centerline to a quarter line (1).

You can also ride it by moving off the wall to some designated point, for example, from the wall to the quarter line (2).

A diagonal line can also be ridden as a leg yield (3).

In each of these examples, the horse's body should always be nearly parallel with the wall and the flexion should be in the opposite direction of travel. The forehand should be slightly ahead of the hindquarters in the sideways part of the movement. In other words, the horse's body is angled ever so slightly in the direction of the sideways part of the travel.

Leg Yield Down the Wall

You can also leg yield down the wall in two different configurations. One is with the nose of the horse to the wall (4), and one is down the wall with the horse's nose facing the inside (5). The easier of the two is the leg yield facing the wall. The wall helps the rider by preventing the horse from escaping the aids, whereas when the horse is facing the inside of the arena, the rider is often inclined to use too much rein to overcome the horse's desire to leave the wall. The angle

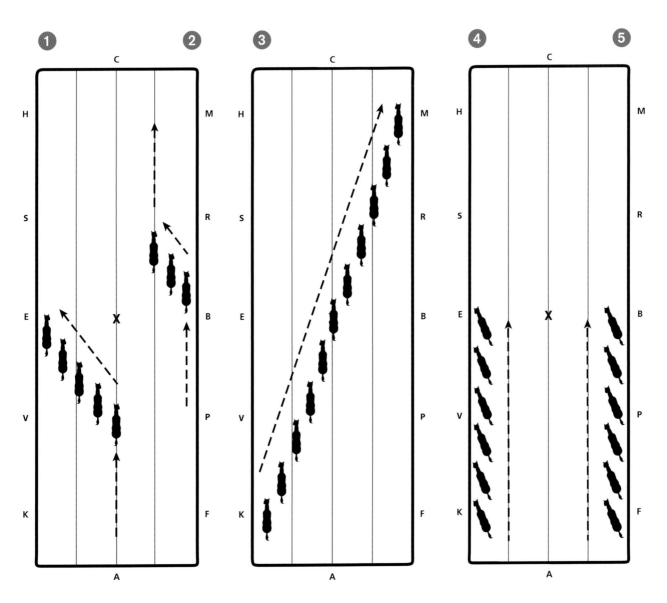

of the horse in each of these examples is in the range of 35 to 45 degrees to the wall.

Spiral In, Spiral Out

The leg yield and the circle can be combined as a spiral for the maximum gymnastic effect from both. The bend in the horse's body is a natural consequence of it matching the arc of the circle. The spiral is ridden by starting on a 20-meter circle and slowly spiraling in to a smaller circle, generally no smaller than 10 meters. At that point, you slowly spiral back out to 20 meters by using leg yield aids to have your horse's inside feet step regularly and evenly in front of the outside feet.

THE LEG YIELD DEBATE

The leg yield is the odd duck within the lateral movements because it does not include bending and collection as do the others. This causes a fair amount of debate among trainers as to whether the leg yield is a "true" lateral movement. Some very good sources keep a leg yield outside the lateral movement category. For example, the German National Equestrian Federation in its book *Advanced Techniques of Dressage*: "Lateral movements are exercises in which the horse steps forward and sideways while maintaining a uniform lateral bend . . . The various forms of leg yielding are not true lateral movements because the horse is only flexed and not bent, and does not need to be collected."

According to the FEI, the leg yield is a lateral movement, but it's different from the others – the purpose of the movement is "to demonstrate the suppleness and *lateral responsiveness* of the horse." The general FEI section on lateral movements says, "The main aim of lateral movements – *except leg yielding* – is to increase the engagement of the hindquarters and thereby also the collection." (Emphasis added.)

In the interest of full disclosure, there is also a degree of debate on the value of doing leg yields. The current assessment of the exercise seems to be divided into two schools of thought. Some trainers barely touch the exercise, insist it is only slightly useful, and say it should be restricted to walk and working trot. On the other side are some highly regarded trainers who school the leg yield in walk, trot, and canter, incorporating it into nearly every training session.

How to Ride a Leg Yield

The aids for leg yield follow the order of your aids for everything else: weight first, then leg, and finally hands. A key element to remember when asking for the leg yield is that "forward" is more important than "sideways."

For a close-up look at how to ride the leg yield, we'll use the example of riding to the wall from a quarter line:

START BY MAKING a correct turn onto the quarter line and confirming that your horse is straight. This is important because you don't want your horse simply slipping through your outside aids and being unbalanced from the beginning.

MAKE A SUBTLE SHIFT in weight to the seat bone that's in the direction you'll want to be going, in this case the outside seat bone.

THE INSIDE LEG should be positioned just behind the girth and will be asking the horse for a forward and sideways movement. The outside leg is in a guarding position behind the girth, keeping the horse from falling toward the wall and the hindquarters from swinging too far.

THE OUTSIDE REIN acts as a regulating rein, keeping the horse from falling through the outside shoulder, but with enough give that the inside flexion is possible. The inside hand is softer and asking for flexion.

▲ A GOOD LEG YIELD

As with all exercises, timing plays an important role. The key moment to ask your horse to step forward and sideways (with an inside leg aid) is when your horse's inside hind leg is just leaving the ground. If you ask at this moment, his inside hind leg steps a little more actively underneath him.

Feeling that exact moment can be tricky. Arena mirrors are helpful, as is having a ground person watch and say *now* when the timing is right or give you feedback on whether you were early or late.

Other routinely seen problems include a horse whose front end is considerably ahead of (or leading) the haunches, or one whose haunches are considerably ahead of the forehand. In either case, the horse is not tracking parallel with the outside wall.

▲ If the haunches are too far ahead of the forehand, as seen here, or vice versa, the horse is not tracking correctly.

Fixing Mistakes in the Leg Yield

I once heard a term that I like: "inside rein-itis," meaning using too much inside rein. It's a disease that runs rampant in riding. Perhaps nowhere does it show up more regularly than when trying to ride a leg yield. It is quite common to see a rider overuse the inside rein in an attempt to get inside flexion.

This leads to an outside shoulder that falls out and a horse that is moving sideways, but in an out-of-balance, sliding-on-ice sort of way. The overflexed horse can also end up being bent around the asking leg, and one of the very clear rules of leg yield is that the horse is flexed but not bent.

Loss of Rhythm

The hands are often to blame for another common error: loss of rhythm. When a rider tries to do too much with the hand, the horse will lose his rhythm. The horse will go against the rider's hand and become tight in the neck. (Remember from the training scale that correct rhythm is the foundation of all riding.)

Rider Error

When trying to diagnose the source of the problems you might be having, remember to look to yourself first. Errors in position of seat, legs, and hands, as well as errors in aids, are quite common when trying to execute a leg yield. The leg yield is where the classic "collapsing of the hip" is seen. This is also where the inside leg can be seen coming too far back and cramping up. (See page 37.)

▲ Too much inside rein leads to too much flexion and a popped outside shoulder.

The horse is flexed in leg yield, but not bent.

CHAPTER 9

THE DESTINATION: INTUITIVE RIDING

After a journey filled with diagonals and half halts, where will your riding take you? Your destination should be a rapport between horse and rider that transcends the mechanics of seats and legs, and of flexion and suppleness, to a connection that's a deep, almost intuitive partnership. The 40 Fundamentals are the stops along the way to the destination of intuitive riding.

INTUITIVE RIDING — or the lack of it — is the reason *why* some people can really ride while others function fine technically but just never seem to fully get it. The question I'm confronted with time and again is "Why, even if a person understands everything about riding theory, does she still seem to fall short of her goals as a rider?" The answer lies in the inner sense of communicating with the horse, and a good name for it is "feel."

This feel might also be described as a sixth sense, a concept that seems to separate the tacticians from the masters in every sport and athletic endeavor. I would make the case that riding a horse requires a more expansive use of the sixth sense than any other sport I know. This is not a reference to some unproven New Age "science" that runs along the lines of ESP. It is a deep connection for the rider who is fully tuned in to her horse's mind and body. Riding, after all, is much more than just a sport or physical activity; it crosses the bridge into artistry.

Figure skating is another example of a sport with a combination of artistry and athleticism that attracts both participant and spectator. But when you bring together the intelligence of a human being with the power and majestic beauty and separate intelligence of a horse, something magical happens. This unique mixture of sport and art, horse and human, sinks deep in the soul of the person riding in partnership and harmony with a horse.

People who have this sixth sense for riding have been called, among other things, "horse whisperers," noted for their abilities to speak the language of the horse, understand what the horse is feeling, and anticipate his response. In return, the rider is able to communicate with and influence the horse in a way that the horse understands.

The rapport between horse and rider is based on an understanding of the nature of each, refined by practice in the saddle, observation of other horses and riders, and time spent just being with horses in the stable and even in the pasture. To understand that rapport requires an understanding of the nature of the horse, the nature of the rider, and how people and horses work together in the rider-horse partnership.

The Nature of the Horse

Much of the horse's nature has been demystified in recent years with the emergence of "natural horsemanship." (In reality, of course, natural horsemanship is nothing new; the ideas have been around for centuries, and probably longer. Much of what Xenophon says about kindness and patience with the horse is surprisingly similar to modern natural horsemanship ideas.) Regardless of what it's called, understanding the nature of the horse is the starting point for intuitive, sixth-sense riding.

Horses by nature are creatures of fight or flight, as they need to be for self-preservation in the wild. They will always choose a quick flight away from danger if escape is at all possible. As potential prey, horses operate with all their instinctual antennae on constant alert for signs of danger. Their mobile ears are constantly picking up sounds that, in a natural

▲ The social nature of the horse is a major factor in their ability to communicate and bond with humans.

setting, would tip them off that a predator is lurking somewhere and preparing to attack.

Their large nostrils constantly take in scents, especially those that are new and warn of a potential danger. When they're scared, they will stop and snort to try to glean more information from what scent might be in the air. The horse's eyes are among the largest of any four-legged creature. These huge eyeballs magnify everything, making it possible for them to see distant objects in clear detail. They have nearly 360-degree vision, again to catch the slightest flicker of danger.

A Social Animal

The herd nature of the horse dictates, without exception, that his daily existence depend on a strong leader. In any size group of horses, a hierarchy will form. This structure is tested every day and sometimes adjusted based on which horse is feeling and

THE SPECIAL NATURE OF THE HUMAN-HORSE CONNECTION

The relationship between the human and the horse is ancient and one of a kind. Humans have ridden and bonded with other animals, elephants and camels for example, but the emotional intensity that embodies the horse and human relationship is yet to be matched.

Although dogs, pigs, sheep, and cattle were domesticated earlier on the timeline of human culture, the domestication of the horse radically altered the landscape. The role the horse has played in the history of humankind has been well documented in writing and in art. Horses gave a new mobility and tool of domination to cultures that used them well. In all of history, no culture using horses has ever fallen to a culture without horses.

Perhaps the best example is the eleventh century domination of the world by the Mongol Empire. This empire, led by Genghis Khan, became the largest the world has ever known, and was created by an army of horsemen whose skill and use of the horse are unparalleled to this day.

▲ The hierarchy of a herd, even a very small one, is tested frequently to make sure there is one strong leader.

acting more dominant. It's important for the survival of the herd as a whole that a leader who shows signs of weakness is shown the door. Without a daily testing of strength and dominance, the herd is vulnerable to the omnipresent dangers of the natural world.

To understand the horse-rider rapport, it is crucial to understand that this instinctual herd-survival behavior extends beyond the horse-to-horse relationship. It includes you. In the world of the horse, the human can become the herd leader. This is where the "gifted" horse person has the inside track.

The rider who becomes a strong and reliable leader, even if the herd is made up of only horse and rider, makes the horse happier, more relaxed, and content. Horses *want* to turn their responsibility for survival over to a leader who protects them from danger and lets them know that everything is going to be okay. And it's your job to assure them that you're up to that task.

Understanding a horse's instinctual fight or flight response and herd needs are the departure points for developing the sense of feel for riding. When you greet a horse at a stall door or pasture gate, you need to walk, talk, and act like a confident, confirmed leader. This in no way means you are to be rough, loud, or angry. It does mean that you should firmly and consistently set parameters for acceptable behavior and conduct.

▲ A rider needs to understand the equine "fight or flight" response. Flight is almost always the horse's first choice when startled.

The Nature of the Rider

We now examine the human and what *our* nature brings to the partnership. We begin with the topic of how the brain works. Commonly referred to as left brain vs. right brain, it's a theory of the functions and structure of how the mind operates. In the study of hemispheric dominance, it's understood that the two different sides of the brain are responsible for two different ways of thinking. Some people's brains are "wired" heavily on one side or the other.

A left-brain orientation usually describes someone who is analytical, methodical, scientific, and logical, as well as uptight and introverted. A person with a right-brain orientation is random, emotional, artistic, holistic, and intuitive, as well as illogical and irrational. Some people are relatively balanced, or "whole-brained," meaning that they are adept in both modes of thinking. Detailed scientific studies tell us that the reality is far more complex and less clear-cut than this explanation, but this simplified version does help you begin to understand that everyone's wiring is unique, and how we learn (and how we learn to ride) is based as much as anything on what's happening in our brain.

The best riders — the intuitive ones — use both halves of their brains. These riders know the fundamentals cold, of course, but they also have an intuition about their horses. They've spent time in the barn grooming their horses, building rapport, and getting a deep understanding of the horse's reactions. They trust themselves when they feel as though a training session is going nowhere and that their horses need a trail ride as a break. They count

on their own intuition that a particular horse needs an extra bit of leg in the far corner, just because he feels better that way.

RIDING BEGINS AT THE STABLE DOOR

I often hear old-time horse people decry the lack of real horsemanship in the up-and-coming ranks of riders. Many modern stables and lesson facilities provide full grooming and tacking services, so that when students, riders, and owners arrive at the barn, their mount is ready to roll. It's a mistake. If a person can't handle herself and the horse she intends to ride in a way that defines who is and is not the leader, she is setting herself up for a difficult and perhaps dangerous ride. When a person takes the time to learn how to properly lead, groom, and tack a horse, she makes herself a better rider.

Many people who have been in the business for years, and who have watched this growing trend of the last decade or two, point out what a disservice this is to the person who wants to ride well. These authorities come from diverse riding disciplines, including hunter/jumpers, eventers, and dressage riders, and all agree that this trend is creating a class of riders who really don't know as much as they should about horse behavior and horsemanship in the true, broad sense of the word.

Thinking vs. Feeling

Ultimately, riding is not about thinking; it's about knowing. It's developing that intuitive relationship with the horse, and while this takes time and effort, it's not ultimately difficult. When a rider takes the time to develop herself into an all-around good horse person, she is learning the language of the horse and developing (often without even being conscious of it) an intuitive sense of how to respond.

When you're thinking, as opposed to knowing, you're acting out a mechanical process. Once you can sense when a horse is getting anxious (because of your broad experiences with the horse), you can respond intuitively, without thinking.

Learning to ride the 40 Fundamentals in this book will confirm the mechanics of riding, but it will also develop your rapport with your horse. Your instructor won't need to tell you when to give with your reins; your horse and your intuition will let you know, and far faster than any instructor can tell you.

It comes down to the tiniest sliver of time between action and reaction. Thinking is too slow of a process for the ideal horse-human interaction. The brain must review all elements of information coming from the five senses (seeing, hearing, smelling, tasting, and touching), put that information into the context of riding theory (knowledge), and produce an immediate, interactive response. The ability to do this adds up to "intuitive" riding or riding with a sixth sense.

The Elements of Intuitive Riding

The following elements must be in place to develop a sixth sense and begin riding intuitively:

THINKING VS. KNOWING. You need to know the fundamentals of riding. If you are struggling with the concepts of correct contact, or are confused about seat aids and half halts, you will be "thinking" through what you heard your instructor say or what you read somewhere to try to get it right. You won't have the attention to feel the horse. Obviously, nothing is better for developing that knowledge than the time spent working through the fundamentals, both on horseback and watching others do it right (and wrong).

CONFIDENCE. You need to feel that a deep and wide comfort zone surrounds you when you ride. There must be an absence of fear. The confidence comes from time in the saddle, but also from being with horses in other ways. The time you spend grooming, for example, makes you more confident in your understanding of your horse's psychology and reactions.

CONTROL OF NEGATIVE EMOTIONS. You need to be able to ride without letting your emotional state rule the day. Negative emotions are anger, fear, anxiety, tenseness, apprehension, disappointment, aggressiveness, frustration, and hostility, among many others. These negative emotions manifest themselves in tight, tense riders who block their horse's movement. The serious-faced rider who is clenching her jaw or grinding her teeth is in conflict, not harmony, with her mount.

RIDING IN THE NOW. Being in the now simply means you are in the present. Horses, unlike humans, don't dwell on yesterday's triumphs or failures, and they don't worry where tomorrow's food is going to come from. They only care about what is happening to them right now. To join your horse in that mental space is actually both rewarding and refreshing. Tuning yourself in to the moment refines your senses and allows you to feel with a deeper sense of the big picture.

Staying in the Present

As human beings, we are conditioned to analyze and search for deeper reasons with critical thinking. Although we don't want to ultimately ignore the reasons things happen, in the moment of our ride we don't want to stay stuck on the thing that happened 10 seconds ago. Consider the domino effect of riding: the longer you stay focused on the mistake, the more errors and problems crop up, and pretty soon the entire ride has disintegrated. If you are attempting to ride a sitting trot on a 20-meter circle, and your horse breaks into canter without you asking, don't worry about the mistake. Fix the stride you are in now and move on. If you don't, you'll not only lose your trot but also the shape of your circle.

The concept of riding in the now needs to be expanded to include what every "now" moment really means in a ride. Think of your time spent in English class learning the difference between past, present, and future tense. Your riding should always be focused on the present tense with a little bit of future. But save your past tense for when you're off your horse or on a walk break. It is important to go back over the elements of a problematic ride so that, next time, you have a game plan or a deeper understanding of why something happened.

Intuitive Riding in Action

If I've persuaded you that intuitive riding should be your goal, don't be intimidated. The good news is that intuitive riding is available to everyone. It's really just a matter of taking the elements of classical riding — the 40 Fundamentals — and adding the approach and attitude of treating your horse as a full riding partner.

In my experience, the recipe is really pretty simple. As with many things in life, it starts with doing your homework, but that homework is the pleasurable experience of riding your horse. And as with a good dessert recipe, it ends with a treat: the satisfaction of rapport between rider and horse. So here's my recipe:

START WITH SOME BOOK THEORY. Reading and understanding the 40 Fundamentals is a great place to start. Learn something about equine behavior and the history of horses and riders. Read magazines and watch videos to see horses in action.

MIX IN PLENTY OF BEING AROUND YOUR HORSE. That means grooming, watching him in the field, even cleaning stalls. Just being around a barn full of horses and their activity will deepen your understanding.

ADD LOTS OF RIDING TIME. This is actual time spent in the saddle. Take the 40 Fundamentals and put them into action, both in the training arena and beyond it, out into the fields and everywhere else you ride. You can do a leg yield on the trail or a working trot up the hill behind your barn. The point is to make the fundamentals part of your riding everywhere. Every experience you have in the saddle is ultimately a good one because experience is the best teacher of all.

AND THE SECRET SAUCE: RELAX AND ENJOY. Realize that your riding is for pleasure and enjoyment. When you're relaxed and having fun, you're more open to feeling what your horse has to say through his behavior and movement.

By the very privilege of getting on a horse's back, we, as riders, owe it to these beautiful animals to do our best for their health and well-being. And as part of that, we owe it to them to know as much as we can about riding, and to listen as carefully as possible to what the horse has to say to us.

Lucky for us, the process of doing that is an immensely happy one. Intuitive riding, based on strong fundamentals, is deeply satisfying for both the horse and the rider.

How the Book and DVD Work Together

How do you learn about riding? As a riding instructor, I believe that learning comes first and foremost from lessons and time in the saddle. But many riders seek a deeper understanding than they can get in a weekly lesson. Where do you get *that*? In my experience, it's primarily from three sources: first, reading about riding, which gives you understanding of key *ideas*; second, seeing photos that show you a key *moment* in riding; and third, watching videos that show you a key *movement*.

As we started to develop the ideas for *40 Fundamentals*, we knew that a carefully written book with many photos and illustrations was imperative. The written word is important, and you need clear images to help show the ideas and concepts. But something was missing.

A still photo shows only a single instant, but the sport of riding is all about movement. We wanted to bring the theoretical concepts to life with pictures that showed the flow of activity and made clear the many subtle movements that make up each moment of riding. In this book and DVD, we've worked hard to integrate ideas, moments, and movements into a single package.

We explored every possible video tool that could shed new light on the subjects being tackled in the book: crane shots, ladders, cables, pulleys, trucks, high angle, low angle, and wide angle, among others. We spent an entire day shooting bird's-eye views from high atop a 40-foot boom, which gave a completely different presentation of the geometry of the horse itself, of schooling figures, gaits, aids, and many other topics.

We show you jumping from a low angle, at normal speed and in slow motion, and in both the arena and the field. We show you the riding aids in close-up and in long shot, in slow motion down to a freeze-frame, and even backwards, once.

Using an extraordinary number of sophisticated video techniques, we filmed the key movements of horse and rider described in the book. As we reviewed the footage, we were excited by how much the slow motion, frame-by-frame editing had to offer. As those moments that you just can't see in ordinary experience become clear, you'll say to yourself, "Now I get it!"

This book and DVD are organized to complement each other. You can read a chapter on any fundamental that you wish to explore, and let the words clarify the topic. Look at the photographs and illustrations to get a clear picture of what is right and wrong. Then pop in the DVD, organized in the same chapter-by-chapter, topic-by-topic outline, and watch that subject come to life with riders demonstrating the before, during, and after of each concept.

If you'd prefer, start with the DVD, and watch what you're interested in. Then read the text about the subject to gain a deeper understanding. Either way, we've worked very hard to give you new insights into the 40 Fundamentals of riding, helping you and your horse attain your goals in a positive, satisfying way.

◄ Sophie was mighty curious about that overhead boom.

A variety of camera angles, from ground level to 40 feet in the air, captured riding theory in action.

— CAST OF CHARACTERS —

▲ **PAGE XI:** Hollie McNeil on **Sophie** (owned by Marilyn Carbin)

▲ **PAGE 26:** Hollie McNeil on **Maya** (owned by Kayla Breithaupt)

▲ **PAGE 7:** Andrea Beukema on **Maddie** (owned by Hollie McNeil)

▲ **PAGE 27:** Alida Durrant on **Cookie** (owned by Sara Ineson)

▲ **PAGE 11:** Kathleen Hayden on **Beckett** (owned by Malinda Raywood)

▲ **PAGE 46:** Hollie McNeil on **Gamine** (owned by Hollie McNeil)

▲ Page 61: Gail Kapiloff on **Ghinger Ale** (owned by Gail Kapiloff)

▲ Page 70: Hollie McNeil on **Kanan** (owned by Dianne Wojtkun)

▲ Page 86: **Dannavirke** (owned by Hollie McNeil)

▲ Page 91: Andrea Beukema on **Nestle** (owned by Johnson and Wales University)

▶ On video only: Andrea Beukema on **Calea** (owned by Amelia Vinsel)

INDEX

Page numbers in *italic* indicate illustrations or photographs; those in **bold** indicate tables or charts.

OTHER STOREY TITLES YOU WILL ENJOY

101 Dressage Exercises for Horse & Rider, by Jec Aristotle Ballou.
Fully diagrammed standard dressage techniques to create a firm foundation for all performance riding.
240 pages. Paper with comb binding. ISBN 978-1-58017-595-1.

101 Jumping Exercises for Horse & Rider, by Linda L. Allen with Dianna R. Dennis.
A must-have workbook that provides logical and consistent series of exercises with easy-to-follow maps and instructions for all riding abilities.
240 pages. Paper with comb binding. ISBN 978-1-58017-465-7.

The Horse Behavior Problem Solver, by Jessica Jahiel.
A friendly, question-and-answer sourcebook to teach readers how to interpret problems and develop workable solutions.
352 pages. Paper. ISBN 978-1-58017-524-1.

The Horse Training Problem Solver, by Jessica Jahiel.
Basic training theory and effective solutions and strategies in a handy question-and-answer format — the third book in a popular series.
416 pages. Paper. ISBN 978-1-58017-686-6.

The Horse Conformation Handbook, by Heather Smith Thomas.
A detailed "tour of the horse," analyzing all aspects of conformation and discussing how variations will affect a horse's performance.
400 pages. Paper. ISBN 978-1-58017-558-6.

Ride the Right Horse, by Yvonne Barteau.
The key to learning the personality of your horse and working with his strengths.
312 pages. Hardcover with jacket. ISBN 978-1-58017-662-0.

The USDF Guide to Dressage, by Jennifer O. Bryant.
The official guide to this classic discipline of horsemanship, from the origins of dressage to the necessary equipment and the fundamentals of training.
352 pages. Hardcover with jacket. ISBN 978-1-58017-529-6.

These and other books from Storey Publishing are available
wherever quality books are sold or by calling 1-800-441-5700.
Visit us at *www.storey.com.*